Ivar Wangensteen

Power System Economics – the Nordic Electricity Market

2nd edition

Copyright © 2012 by
Vigmostad & Bjørke AS
All Rights Reserved

First Edition 2007
Second Edition 2012 / Printing 3, 2019

ISBN: 978-82-519-2863-2

Grafic production: John Grieg, Bergen
Layout: The author
Cover design: The publisher

Enquiries about this text can be directed to:
Fagbokforlaget
Kanalveien 51
5068 Bergen
Tel.: 55 38 88 00 Fax: 55 38 88 01
e-mail: fagbokforlaget@fagbokforlaget.no
www.fagbokforlaget.no

All rights reserved. No part of this publication may be reproduced, stored in a retrieval system, or transmitted, in any form or by any means, electronic, mechanical, photo-copying, recording, or otherwise, without the prior written permission of the publisher.

Preface the second edition

There are only few changes in the present book compared to the first edition. The most important ones are the inclusion of resent developments in the Nordic market and in other parts of the world. That affects in particular Chapter 5. Chapter 6, 7 and 8 are also revised to take resent changes into account.

In addition to this, some errors in the first edition are corrected.

I am indebted to numerous students at NTNU and at the Indian Institute of Technology (IIT) in Mumbai for feedback on the first edition.

Trondheim December 2011

Ivar Wangensteen

Preface to the first edition

This book is written as a textbook for students of engineering at the Norwegian University of Science and Technology (NTNU). It is designed for the *Power Markets* course which is part of the *Energy and environment* master's programme and the recently established international MSc programme in *Electric Power Engineering*. The book is based on the lecture notes used in this course for some years. Students from other programmes at NTNU also participate in this course.

It is my hope that the book can also be used by other universities, particularly in the Nordic countries.

As the title indicates, it deals with both power system economics in general and the practical implementation and experience from the Nordic market.

In principle, this book requires no previous knowledge of economics although some background would be advantageous. One chapter on basic microeconomics is included to outline some of the basic concepts. Part of the material requires basic knowledge in power engineering, in particular load flow analysis. But except for this, previous engineering knowledge is not really necessary.

Although it is written primarily as a textbook for students, readers outside the universities may also find the book interesting. It deals with problems that have been subject of considerable attention in the power sector for some years and it addresses issues that are still relevant and important.

Acknowledgements:

Several colleagues at NTNU and SINTEF Energy Research have contributed to the content of this book: Arne Haugstad, Gerard Doorman, Anders Gjelsvik, Nicolai Feilberg, Ove S. Grande, Birger Mo, Bjørn H. Bakken, Olav B. Fosso, Hans H. Faanes, Arne Johannesen. Important contributions also come from Richard Christie, University of Washington. Some PhD and MSc students at NTNU have also been involved: Audun Botterud, Anders Kringstad, Karl Magnus Maribu, Tarjei Kristiansen, Ivan Andročec, Ingrid S. Kristensen.

Astrid Lundquist, SINTEF Energy Research is responsible for the illustrations.

Stewart Clark, NTNU has been the language consultant.

I am indebted to all these contributors.

Trondheim December 2006

Ivar Wangensteen

Table of contents

1 Introduction 9
1.1 Short summary 9
1.2 The power system 10
1.3 Traditional organization 11
1.4 Restructuring 11
1.5 Special features of electricity 12

2 Basic microeconomics 15
2.1 Introduction 15
2.2 Supply, demand and the market balance 15
2.3 Demand elasticity 17
2.4 Production cost 19
2.5 Aggregate supply 22
2.6 Competition 23
2.7 Producer and consumer surplus 24
2.8 Imperfect competition (market power) 25
2.9 Externalities 30
2.10 References 32

3 Electricity consumption 33
3.1 General 33
    3.1.1 Characterization 33
    3.1.2 Categorization 34
3.2 Demand drivers 35
    3.2.1 Income elasticity 35
    3.2.2 Price elasticity 38
3.3 Load research 41
    3.3.1 Load measurements 43
    3.3.2 Factors influencing the load level 43
    3.3.3 Load aggregation 44
3.4 Load management 59

3.5 References 60

4 Pricing 61
4.1 Introduction 61
4.2 Marginal cost pricing 62
4.3 Dynamic pricing 66
    4.3.1 Time Of Use (TOU) rates 66
    4.3.2 Spot pricing 70
4.4 Ramsey pricing 71
4.5 Tariffs 75
4.6 References 76

5 Restructuring/Deregulation of the Electricity Supply Industry 77
5.1 Terminology 77
5.2 Economy of Scale and Unbundling in the Power System 77
5.3 Privatization 79
5.4 Stranded costs 79
5.5 Roles and responsibilities in a restructured system 80
    5.5.1 Regulator 80
    5.5.2 Producers 81
    5.5.3 Grid Companies 81
    5.5.4 End users 82
    5.5.5 Power Exchange (PX) 83
    5.5.6 System Operator (SO) 83
    5.5.7 Balance responsible entities 84
    5.5.8 Suppliers 85
    5.5.9 Traders and Brokers 85
    5.5.10 Market Participants 85
5.6 Market design 86
    5.6.1 Market access 86
    5.6.2 Exchange/pooling 87
    5.6.3 Reserves and balancing 91
    5.6.4 Physical and financial trade 92
5.7 Restructuring in different countries 93
5.8 The Norwegian and Nordic restructuring 97
    5.8.1 Introduction 97

5.8.2	Organizational and technical factors	97
5.8.3	Organizational structure before restructuring	98
5.8.4	Major arguments for restructuring	99
5.8.5	The Nordic generating system	100
5.8.6	The organization of the new market	101
5.8.7	The Exchange - Nord Pool Spot and NASDAQ OMX	103

5.9 Developments in the Norwegian and Nordic market 106
- 5.9.1 The pool's influence on the general price level 106
- 5.9.2 Investments and power balance 110
- 5.9.3 Market Efficiency 111
- 5.9.4 Market growth at Nord Pool 113
- 5.9.5 Market information and statistics 114

5.10 Retail access 114

5.11 References 116

6 Power generation 119

6.1 Introduction 119

6.2 Characteristics of generating units 119
- 6.2.1 Thermal units 119
- 6.2.2 Hydropower units 121

6.3 Generation scheduling 123
- 6.3.1 Merit order 124
- 6.3.2 Optimal Dispatch 124
- 6.3.3 Unit commitment 127
- 6.3.4 Scheduling 129
- 6.3.5 Hydro scheduling 130

6.4 Market modeling (EMPS) 142
- 6.4.1 Fundamental models 142
- 6.4.2 The EMPS simulation part 144

6.5 Generation planning in an open market 148
- 6.5.1 Introduction 148
- 6.5.2 Problem formulation in the traditional and deregulated environment 149
- 6.5.3 Unit commitment in an open market 150
- 6.5.4 Hydro generation 152

6.6 Generation planning in practice 157
- 6.6.1 Overview 157
- 6.6.2 Long-term scheduling 159
- 6.6.3 The medium-term scheduling model 161
- 6.6.4 The short-term scheduling model 161
- 6.6.5 Details about the bidding and price clearing process 162

Table of contents

  6.6.6 The Balancing Market 165
6.7 **Optimal mix of generating units.** 167
  6.7.1 Optimal mix based on cost minimization 167
  6.7.2 The revenue problem 170
  6.7.3 Introduction of load shedding 172
6.8 **References** 174

7 Grid access 177

7.1 **Introduction/definition of concepts** 177
7.2 **Congestion Management** 182
  7.2.1 Introduction 182
  7.2.2 Congestion Management Based on Area Pricing 183
  7.2.3 Consequences of a Buy-Back Procedure 194
7.3 **Price Calculation Based on Optimal Power Flow (OPF)** 198
  7.3.1 Principle 199
  7.3.2 Nodal Prices 206
  7.3.3 Marginal transmission losses and nodal prices 208
  7.3.4 Timing and data 209
  7.3.5 Compensation for transmission losses 210
  7.3.6 Location of the market place (hub) 211
  7.3.7 Security of supply 213
  7.3.8 Three-node DC example 216
  7.3.9 Three-node DC Equivalent 220
  7.3.10 Hydro and Hydro-thermal Systems 225
  7.3.11 DC–equivalent. 11 node (zone) power system 232
7.4 **Transmission/Distribution (T/D) tariffs** 241
  7.4.1 General requirements 241
  7.4.2 The Nordic T/D tariff system 242
  7.4.3 Congestion management in the Nordic system 248
  7.4.4 Congestion management in Europe 248
7.5 **Nordic power exchange before 1991** 251
  7.5.1 Responsibility for operation of the Nordic system 252
  7.5.2 Power exchange in case of disturbances 253
7.6 **References** 255

8 Ancillary services 275

8.1 **Purpose and definition** 275
8.2 **Classification** 277
  8.2.1 Balancing services 281

		8.2.2 System services 281
8.3	Reserve requirement 282	
8.4	Automatic control 283	
8.5	Balancing 287	
8.6	Costs 289	
	8.6.1 Costs for Active Reserves 289	
8.7	A simple model for pricing of reserves 292	
	8.7.1 Purpose 292	
	8.7.2 Assumptions and border conditions 292	
	8.7.3 Balance between the Spot Market and the reserves market 293	
8.8	The Norwegian and Nordic system 297	
	8.8.1 Primary reserve. 297	
8.9	Some experiences 299	
	8.9.1 The Balancing Market (BM) 299	
	8.9.2 The Reserve Option Market (ROM) 301	
	8.9.3 Special conditions for consumption side reserves 304	
	8.9.4 Results from the first two rounds 305	
	8.9.5 Development during the next few years 307	
8.10	References 308	

9 Costs and regulation of grid monopolies 311

9.1	Grid costs 311
9.2	Economy of scale in electricity distribution and transmission. 312
9.3	Monopoly regulation 315
	9.3.1 Definition 315
	9.3.2 Pricing for a natural monopoly 315
	9.3.3 The objective for monopoly regulation 316
	9.3.4 Presumptions 318
	9.3.5 Alternative mechanisms 319
9.4	The English price cap regulation 323
9.5	The Norwegian revenue cap regulation 324
	9.5.1 Overview 325
	9.5.2 Price indexing 326
	9.5.3 Efficiency requirement 327
	9.5.4 Efficiency measurement by DEA 327
	9.5.5 Revenue to cover costs for increased delivery 329
	9.5.6 Compensation for Energy Not Supplied (CENS) 329

9.6 References 331

10 Environment policy 333
10.1 Introduction 333
10.2 Policy instruments, overview 334
    10.2.1 General instruments 334
    10.2.2 Sector specific instruments 335
10.3 Cap or tax 337
10.4 Environmental taxes and subsidies 339
    10.4.1 General 339
    10.4.2 Operational consequences 341
    10.4.3 Long term impacts 342
10.5 Cap and trade 345
10.6 Tradable Green Certificates 347
    10.6.1 The goal of TGCs 348
    10.6.2 Supply and demand of green certificates 349
10.7 TGCs and standard economic theory 350
    10.7.1 A socio-economic approach 351
    10.7.2 Conditions for a stable TGC market 361
10.8 References 362

Abbreviations 363

1 Introduction

1.1 Short summary

This book focuses on power system economics with emphasis on restructured or deregulated[1] power supply systems. That means a power system where generation and sales are subject to competition. Prior to deregulation, the power supply was undertaken by utilities operating as franchised monopolies. Their objective was to supply consumers with electricity on a minimum cost basis. This book also considers economic operation of the power system under these former circumstances.

The power system includes both a supply and a demand side. In traditional power system engineering, attention was concentrated on the power supply side. The link to demand was through the tariffs. Operation and investment planning with extensive use of complex optimization tools has been – and is still - a power engineering specialty. We will not go into detail about these optimization algorithms in this book. Only a short overview is presented. However, considerable attention is given to how these tools had to be adapted to the new situation and how they are used after deregulation. (Chapter 6).

The restructuring of the electricity supply industry brought market economy into the power sector. The reform focused on how to bring competition into the electricity supply industry. A short introductory description of some basic microeconomic concepts is given in Chapter 2. It is acknowledged however, that parts of the supply system, i.e. the grids, are natural monopolies and have to be treated accordingly. The restructuring process therefore included an unbundling of the system whereby the competitive and monopolistic activities were split.

The restructuring of the power sector entailed new planning criteria. In Chapter 6 it is described how the traditional planning criteria and planning tools for power generation had to be adapted to this new situation.

For the grid companies economic regulation was introduced. Chapter 9 describes some general aspects of economic regulation and the Norwegian revenue cap regulation is summarized.

General theories as well as practical applications are described in this book. The description of practical solutions and experience are to a large extent taken from the Norwegian and Nordic power systems, but there is also relevant information from other parts of the world.

---

[1] Restructuring and deregulation are here used synonymously here.

Chapter 1: Introduction

1.2 The power system

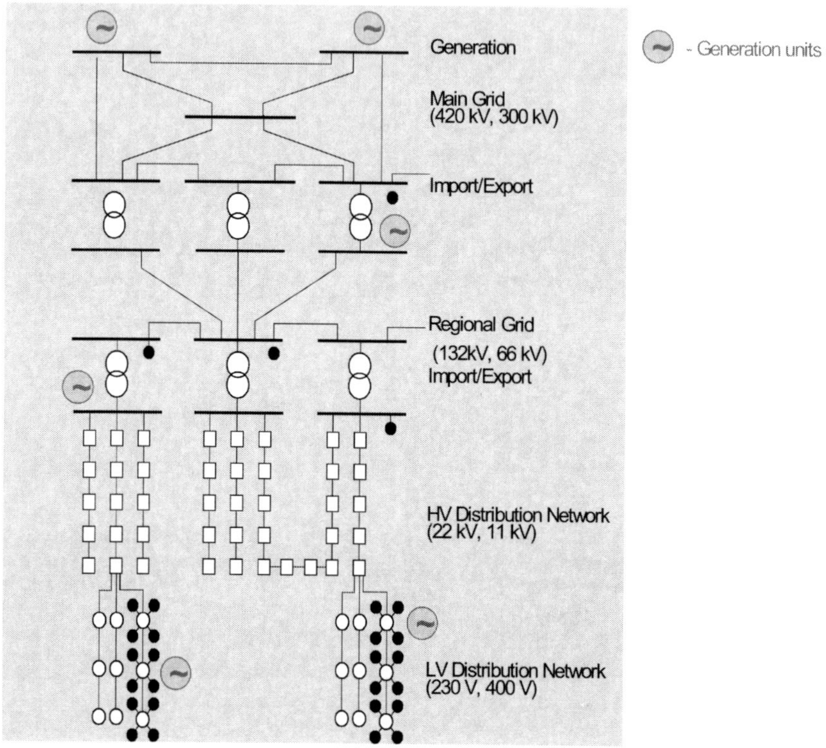

Figure 1.1 The power system (Norwegian case).

Figure 1.1 gives an overview of the power system "hardware". Notice that the power system consists of both a demand side and a supply side. Traditionally the consumers have been described as "non dispatchable". They can be affected by pricing, or more generally by tariffs, but not by direct control.[2]

On the supply side we distinguish between generation and transmission/ distribution (T/D).

The whole system is physically interconnected. A minor action in one part of the system will in principle affect the whole system. This interdependency can be described precisely by physical laws.

---

[2] Modern information and communication technology offers improved possibilities for direct control of individual end users. Combined with advanced tariffs this can be used to monitor and control consumption and thereby improve the efficiency of the system.

1.3 Traditional organization

A brief look at how the electricity supply industry was organized before the deregulation trend started about 1990, reveals two typical patterns:

- *Privately owned utilities with public regulation.* This was typical for the USA. Electricity supply was (and is still) dominated by Investor Owned Utilities (IOUs), which was regulated by public regulatory commissions. The regulatory commissions had a decisive impact on tariffs.[3]
- *Publicly owned utilities.* This was typical for Europe. We find either centralized State owned utilities as in France (EDF), England (CEGB) and Italy (ENEL), or we find a decentralized utility pattern based on the State, county and municipal ownership as in Scandinavia.

In a few cases we find a mix between the two patterns where utilities are dominated by private ownership, but not subject to any direct public regulation. This was the standard in Spain and Germany. However, these countries had some public control, although not in the same direct and visible way as in the USA.

Common for all the countries was that the utilities had a monopolistic position in the area they served, i.e. they were franchised monopolies implying that they had a concession to operate as monopolies within a defined area. The consumers could only buy from one supplier and there was little room for Independent Power Producers (IPPs). In some cases, however, IPPs were allowed and even encouraged to participate at an early stage. A good example is the USA, where the Public Regulatory Policies Act (PURPA) from 1978, gave the utilities a legal obligation to buy electricity from IPPs at prices based on "avoided cost". This legislation gave good opportunities for IPPs.

1.4 Restructuring

The pioneer in European restructuring was England & Wales, where privatization and deregulation came with the Electricity Act of 1989. Norway followed with the Energy Act of 1990 and the other Scandinavian countries and Finland joined this market during the 1990s (more about the Nordic development below). Spain (1998) and the Netherlands (1999) have also created fully competitive markets. Germany has introduced full retail access to the grid, but certain elements remain before the German market is fully competitive.

The restructuring process in the European Union is partly a result of these early national initiatives and partly a result of initiatives from the European

---

[3] A large part of the rural areas in the USA are covered by cooperatives, i.e. customer owned utilities which are not subject to public regulation. However, the number of customers served by cooperatives is small compared to the number served by IOUs.

Commission. Directive 96/92/EC, December 1996 supplemented by Directive 2003/54/EC from June 2003 are aiming at a fully open electricity market by 2007. One interesting observation is that the last directive (2003/54/EC) emphasizes the responsibility of the member states for ensuring a high level of security of supply. This concerns network security as well as maintaining sufficient generating capacity.

The restructuring is described in more detail in Chapter 5.

1.5 Special features of electricity

Electricity has certain features that make it a unique commodity.

The following list captures the essentials:

- *Continuous flow.* Electricity is generated and consumed continuously. Gas transported through a gas grid has basically the same feature.

- *Instant generation and consumption.* Electricity is consumed at the same moment of time as it is generated. If we compare it with gas, the transport speed of gas in a pipe is about one metre per second. Electricity travels at the speed of light.

- *Non-storability.* Electricity cannot be stored in significant quantities in an economic manner.

- *Consumption variability.* Electricity consumption is variable with a characteristic pattern during day/night, during a week and a year.

- *Non-traceability.* There is no physical means by which a unit of electricity (a kWh) delivered to a consumer can be traced back to the producer that actually generated the unit. This feature puts special requirements on the metering and billing system for electricity.

- *Essential to the community.* Electricity is regarded as an absolute necessity in modern society. Practically every household and every firm has a connection to the power grid. How essential electricity is can be illustrated by the Value of Lost Load (VOLL)[4], which is sometimes estimated to be 100 times higher than the ordinary price of electricity.

- *Breakdown possibility.* Due to the technical characteristics of a power supply system, complete system breakdown can occur. Large areas can be affected. We have seen large breakdowns, for instance in New York in 1977 and 2003. Italy had a complete blackout in September 2003. The financial consequences of such incidents are tremendous.

---

[4] VOLL is the loss to society (consumers, firms) caused by a sudden interruption of the electricity supply.

Special features of electricity

In all commodity markets there must be balance between production and consumption over a period of time. In the electricity market there must be an <u>instant</u> balance. Generation and consumption have to balance, minute-by-minute, day and night throughout the whole year. Economic theory is based on simultaneous balance between supply and demand, and that balance is created by the price, which consumers and producers observe and adapt to. However, in the power system the extreme simultaneousness and the continuous load variability create a problem. The price mechanism cannot work fast enough to balance generation and consumption in real time. One practical consequence is that electricity pricing always has to be either ahead of real time, <u>ex ante</u>, or after real time, <u>ex post</u>. Strictly speaking, there can be no real-time market for electricity.

The real-time balancing mechanism in a power system is based on frequency, not on price. If a an imbalance occurs, i.e. due to the outage of a generating unit, a drop in frequency follows, which leads to increased output from the remaining units caused by the automatic frequency control. At the same time a slight drop in the load occurs. In that way the generation/consumption balance is maintained. We return to that in more detail in Chapter 8.

One interesting aspect of a power system is the possibility of describing it in mathematical terms based on physical characteristics. This is because it is an interconnected physical system. That makes it possible to model the economic performance of the electricity supply system in more detail than is possible in other economic sectors. This special feature, combined with the fact that it is an extremely extensive and complex system makes the power system a very interesting and challenging area for operations research. There is a long tradition for the use of complex optimization tools for planning of operation and expansion of the power system.

2 Basic microeconomics

2.1 Introduction

The purpose of this chapter is to give an introductory description to some basic microeconomic concepts, which are frequently made use of in the following chapters. This chapter will only be of interest for those not familiar with economic theory.

The description is very short. There are several textbooks giving a more comprehensive description. One of them is included in the reference list [1].

2.2 Supply, demand and the market balance

The supply of a commodity, i.e. production and marketing, increases as the market price rises. This is because the marginal cost normally increases with increasing quantities. On the other hand, the demand for a commodity, i.e. the quantity consumers are willing to buy, normally decreases when there is a price rice. This is because the usefulness (in economic terms: the *utility*) of a marginal increase in the consumption, usually decreases with increasing use of the commodity. The market balance is determined by the supply-demand equilibrium as shown in Figure 2.1.

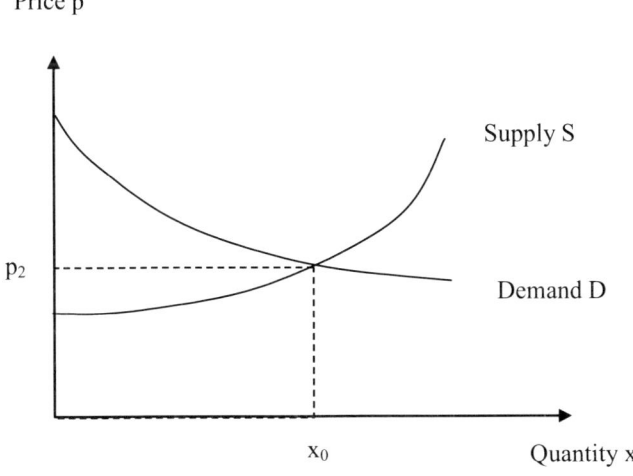

Figure 2.1 The Marshallian Supply-Demand Cross.

The supply-demand cross is fundamental in economic theory. It was introduced by the English economist Alfred Marshall (1842 – 1924), who found that supply and demand interact to determine the market balance and thereby deciding the price and the quantity. The market balance is represented by the point where the supply and demand curves intersect, as shown in the figure. This point is an *equilibrium* where both buyers and sellers are content with the quantity being traded and with the price at which it is traded.

Another important aspect in classical economic theory is that supply and demand operate *simultaneously* to determine the price. The practical interpretation of this is that the trading process, whether it is an auction or a set of bilateral bargaining processes, leads to one market price. This price is seen by producers and consumers during a certain time span, and gives them sufficient time to adapt to the observed price.

This state of *simultaneous equilibrium* is important in economic theory. The economy is normally assumed to be in a state of equilibrium. In this state there is only one price for a given commodity at a given time. With a perfect market, this state of equilibrium represents an optimal solution to society, implying maximum economic efficiency, also denoted maximum social welfare. For one single commodity it is called a *partial equilibrium*. If all commodities are included we talk about a *general equilibrium* and a general societal optimum.

Optimality as it is defined in that context is a Pareto[1] optimality, which means that no person can be made better off without someone else being made worse off. There are two main theorems in welfare economics: 1) a competitive equilibrium is a Pareto optimum, and 2) any Pareto optimum can be sustained as a competitive equilibrium [2].

Pareto optimality does not include any consideration of distributional aspects. This means that the distribution of a society's wealth can be skewed, but it is still Pareto efficient[2].

An important characteristic of this competitive equilibrium is that no mutually beneficial trading opportunities are unexploited. As a consequence there is no arbitrage between different markets, different locations (locational) and different instants of time (temporal).

---

[1] After the Italian scientist Vilfredo Pareto (1878 – 1923)
[2] It can be argued that it is misleading to use the term "welfare maximum" or even "social welfare maximum" in the case where no distributional effects are taken into account. However, this is a well-established terminology in economics and it is also used in this book, although in most cases the term "economic efficiency" is used. The meaning is essentially the same.

2.3 Demand elasticity

With a few exceptions[3], the demand for a certain product declines with increasing price as indicated in Figure 2.1. Based on this downward-sloping demand curve one can define *demand elasticity*, which describes how demand depends on price.

We start with a simple function:

$$x = f(p) \qquad (2.1)$$

describing the demanded quantity x as a function of the price p. A small increase of the price Δp leads to a corresponding change Δx in quantity.

$$\Delta x = \frac{\partial f}{\partial p} \Delta p \qquad (2.2)$$

If we want to find the relative (or percentage) change in quantity as a result of a relative (or percentage) change in price, we can write Equation (2.2) as:

$$\frac{\Delta x}{x} = \left(\frac{\partial f}{\partial p} \frac{p}{x}\right) \frac{\Delta p}{p} \qquad (2.3)$$

and the factor:

$$e = \left(\frac{\partial f}{\partial p} \frac{p}{x}\right) \qquad (2.4)$$

is defined as the elasticity of x with respect to p. If the elasticity is $e = -0.5$, it means that one per cent increase in price gives 0.5 per cent decrease in quantity.

---

[3] One exception is the so-called "Giffins' case". It was observed that the demand for potatoes in Ireland increased with rising price. This was explained by the fact that potato had a large market share and was typical cheap food. The increased price of potatoes led to greater poverty, which in turn caused even more use of potatoes. Another example is the so-called "Veblen effect" (named after the economist/sociologist Thorstein Veblen), which applies to typical luxury goods. The demand for such commodities can increase with rising price due to the prestige of using an expensive product.

The demand for a product does not depend on the price of this product alone. It depends on the price of all products, p_1, p_2, p_3, ... p_n and the income, I. Mathematically it can be expressed as n demand functions of the form:

$$x_1 = f_1(p_1, p_2, p_3, ... p_n, I)$$
$$x_2 = f_2(p_1, p_2, p_3, ... p_n, I)$$
$$\vdots$$
$$x_n = f_n(p_1, p_2, p_3, ... p_n, I)$$

(2.5)

Corresponding to this we can define the *direct price elasticity*, which is normally negative, as mentioned before:

$$e_{11} = \frac{\partial x_1}{\partial p_1} \frac{p_1}{x_1}$$

(2.6)

In addition we can define a *cross price elasticity*:

$$e_{12} = \frac{\partial x_1}{\partial p_2} \frac{p_2}{x_1}$$

(2.7)

The cross price elasticity, e_{12}, is the percentage change in the demand of product 1 when the price of product 2 increases 1 per cent. If the cross price elasticity is positive, the corresponding products are substitutes (i.e. the demand for product 1 goes up when the demand for product 2 goes down). If the cross price elasticity is negative, the corresponding products are complements (i.e. the demand for product 1 goes up when the demand for product 2 goes up).

In addition to prices, the income will have an impact on the demand for a product. The income elasticity is defined as:

$$E_1 = \frac{\partial x_1}{dI} \frac{I}{x_1}$$

(2.8)

The income elasticity is normally positive (i.e. a rise in income gives increased demand), although it in some cases can be negative.

A market demand curve is the "horizontal sum" of each individual demand curve. For each price the total quantity demanded by the market is the sum of the amounts demanded by each individual. This is shown in Figure 2.2 at a given price, e.g. p*, the market demand is:

$$x(p^*) = x_1(p^*) + x_2(p^*) \qquad (2.9)$$

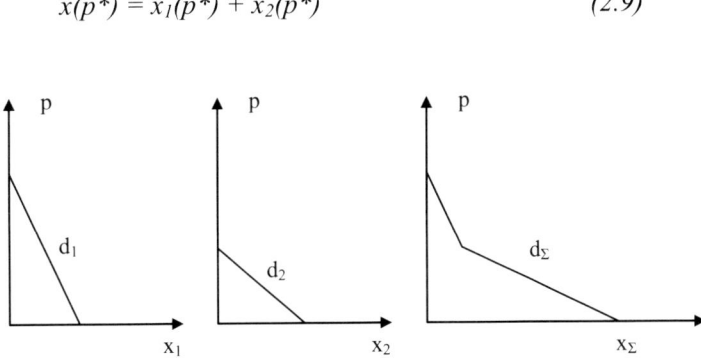

Figure 2.2. Construction of a market demand curve from the individual demand curves.

2.4 Production cost

The production cost as a function of quantity (output per period of time) for a given firm can be as shown in Figure 2.3. This shape is typical for short-term variation of cost as function of short-term variation of output. There is a fixed cost component and the total cost, C, generally increases with increasing output until a production limit is reached.

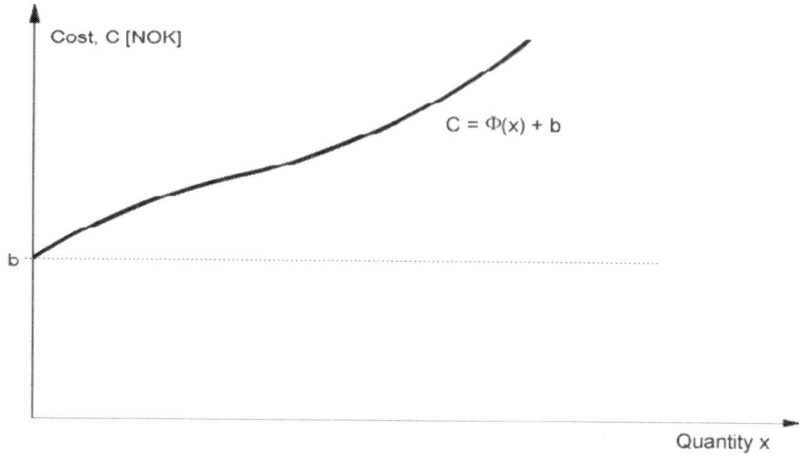

Figure 2.3 Production cost.

Mathematically the production cost can be formulated as:

$$C = \Phi(x) + b \qquad (2.10)$$

Figure 2.4 shows different unit cost curves and their corresponding mathematical expressions are as follows:

Average Total Cost:

$$ATC = \frac{\Phi(x)+b}{x} \qquad (2.11)$$

Average Variable Cost:

$$AVC = \frac{\Phi(x)}{x} \qquad (2.12)$$

Average Fixed Cost:

$$AFC = \frac{b}{x} \qquad (2.13)$$

Marginal Cost:

$$MC = \frac{d\Phi(x)}{dx} \quad (2.14)$$

It can be shown that the marginal cost curve always passes through the point of minimum average variable cost, AVC, and the point of minimum average total cost, ATC.

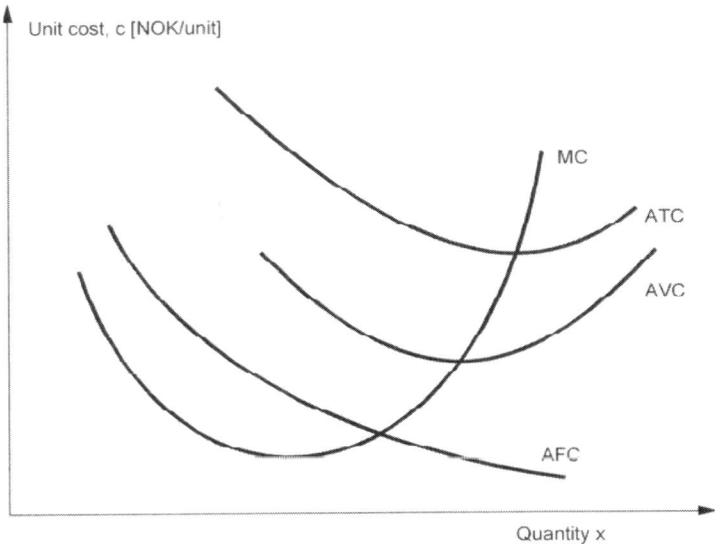

Figure 2.4 Short-run costs per unit

The cost curves in Figure 2.3 and Figure 2.4 are all drawn with the assumption that no new investments are done. This means that they are short-run cost curves.

If we invest in new production facilities, we get a new short-term cost curve. The fixed cost is higher, but with increasing output it reaches a point where the total cost becomes lower than with the original cost curve. We can construct a set of different short-run cost curves for alternative levels of investment, as shown in Figure 2.5.

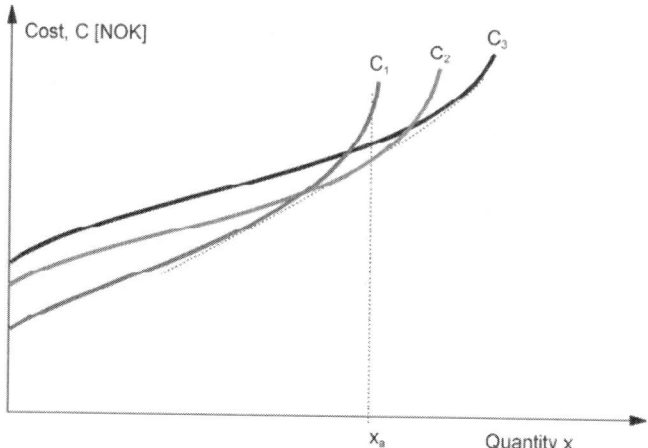

Figure 2.5 Production cost for different levels of investment.

We can now construct an enveloping curve, as shown in the Figure 2.5. The enveloping curve shows how it is possible to increase output at minimum cost by making new investments. This is the long-run cost curve for this product. The corresponding derivative,

$$LRMC = \frac{dC_{LR}(x)}{dx} \qquad (2.15)$$

is the long-run marginal cost.

For the output level x_a indicated in Figure 2.5 we see that C_2 represents the lowest cost and is thus the optimal investment level. C_1 has a higher cost for this level of output, indicating an under-investment. C_3 has also a higher cost but in this case it represents an over investment. At the output level x_a on the C_2 curve we see that the long-run marginal cost equals the short-run marginal cost (or LRMC = SRMC), which is a general rule for optimal investments.

2.5 Aggregate supply

A profit-seeking company will increase its output as long as the price is higher than the cost of producing one additional unit. Therefore, the short-run marginal cost curve is also the short-run supply curve for a producer, provided the company is a *price taker*. However, in addition to the short-run marginal cost, a production company will take the average variable cost into account. It will not produce anything until it is able to cover its variable cost. Consequently, the supply curve for a producer will be the part of the short-run marginal cost curve that is higher than the AVC-curve.

For a set of competing producers in the market, we can construct an aggregate supply curve by adding the marginal cost curves of each individual firm "horizontally". This is shown in Figure 2.6.

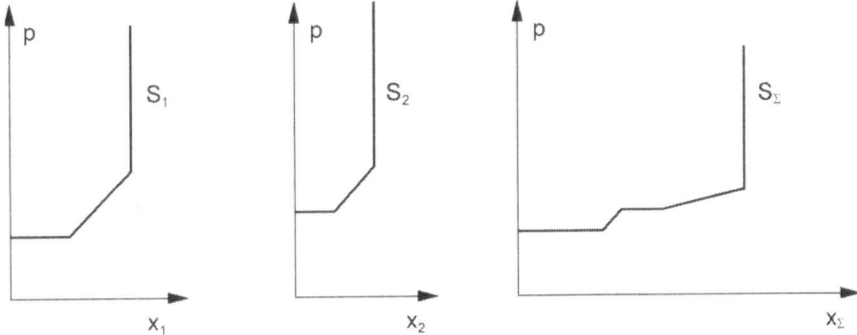

Figure 2.6 Horizontal addition of marginal cost curves into one resulting supply curve.

2.6 Competition

As stated in Section 2.2 a competitive market, or more precisely: a perfectly competitive market, leads to maximum economic efficiency. There is a set of conditions that have to be met in order to obtain a perfectly competitive market:

- Each market participant must be too small to be able to affect the market price, i.e. all market participants must be *price takers* (Section 2.8 describes the consequences if this condition is not met).
- All market participants have to be economically rational, i.e. producers maximize their profits and consumers maximize their utility.
- All market participants should have *perfect knowledge* about prices and other factors of importance for their decisions.
- There has to be *free entry* to the market for new producing firms.
- Products and production factors should be traded freely and with no transaction cost.

All these conditions are never fully met in the real world, but we can be reasonably close.

We can broadly distinguish between two sources of inefficiency. The first is the so-called *x-inefficiency*, which is internal inefficiency in the production companies, like old technology, poor management and overstaffing. The second is *market inefficiency*, which means that the price does not reflect marginal cost and marginal willingness to pay.

A competitive market will create incentives to avoid both these types of inefficiency. A company with a profit-maximizing objective will obviously have an incentive to avoid x-inefficiency simply because it reduces profit. If it this not done, the company can be out of business. It is also easy to show that market inefficiency will be avoided:

We assume the producer to be a price taker and has the following short-term production cost.

$$C = \Phi(x) + b \qquad (2.16)$$

The producer's profit will be:

$$\Pi = xp_x - (\Phi(x) + b) \qquad (2.17)$$

Profit maximizing behaviour implies that:

$$\frac{d\pi}{dx} = 0 \qquad (2.18)$$

which gives :

$$p_x = \frac{d\phi(x)}{dx} \qquad (2.19)$$

The result is that that the price should equal Short Term Marginal Cost (STMC), which leads to maximum economic surplus as we shall see next.

2.7 Producer and consumer surplus

The shaded areas in Figure 2.7 represent producer and consumer surplus.

Imperfect competition (market power)

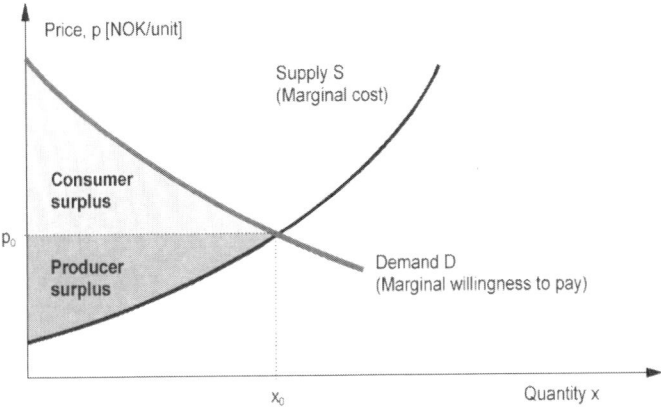

Figure 2.7 Producer and consumer surplus.

The total economic surplus to society, i.e. the social welfare, is represented by the sum of producer and consumer surplus. The total surplus is maximized if the supply and demand balances by a price equal to marginal cost (*SRMC*). Because the supply curve reflects the SRMC of the producers, the area under the supply curve represents the variable cost for the operation of the existing production equipment. This producer surplus can partly be used to cover the fixed cost for the production system, including interest and depreciation on the invested capital.

This optimal solution represented by the price cross can be obtained by a perfectly competitive market. But in order to be an optimal solution to society, it is required that there should be no *external costs*. We return to this in Section 2.9. Another source of inefficiency is *imperfect competition or market power*.

2.8 Imperfect competition (market power)

There are different forms and degrees of imperfect competition. If a participant on the market is large enough to affect the market price by its actions, it can increase its profit (or its utility if the participant is on the demand side). By doing so the participant causes a loss in economic efficiency. Here, we will start with a description of the simplest version of imperfect competition; *monopoly*[4]. Next we will present a model for an *oligopoly*.

---

[4] We are here considering a company with a monopolistic position on the supply side. But we can also have monopolies on the demand side. In that case we call this *monopsony*.

2.8.1 Monopoly

If a producer is in a monopolistic position on the market, it is able to control the price completely by offering different quantities to the market. The producer is no longer a *price taker*, but a *price maker*. The producer will not regard the price as fixed, but as a function of the quantity produced.

$$p_x = p_x(x) \qquad (2.20)$$

The profit expression will not change:

$$\Pi = xp_x(x) - (\Phi(x)+b) \qquad (2.21)$$

However, in this case the condition for profit maximization,

$$\frac{d\pi}{dx} = 0 \qquad (2.22)$$

will be different,

$$x\frac{dp_x}{dx} + p_x - \frac{d\phi}{dx} = 0 \qquad (2.23)$$

leading to

$$\frac{d\phi}{dx} = p_x + x\frac{dp_x}{dx} \qquad (2.24)$$

The right hand side of Equation (2.24): $p_x + xdp_x/dx$, is the Marginal Revenue (MR) for the monopolistic producer. The last part, xdp_x/dx, is normally negative, due to the fact that price (or willingness to pay) is declining with increased quantity. This means that the optimal market price, p_x, is higher than the marginal cost. In Figure 2.8 a monopoly situation (x_2,p_2) is compared to a fully competitive market (x_1,p_1).

Imperfect competition (market power)

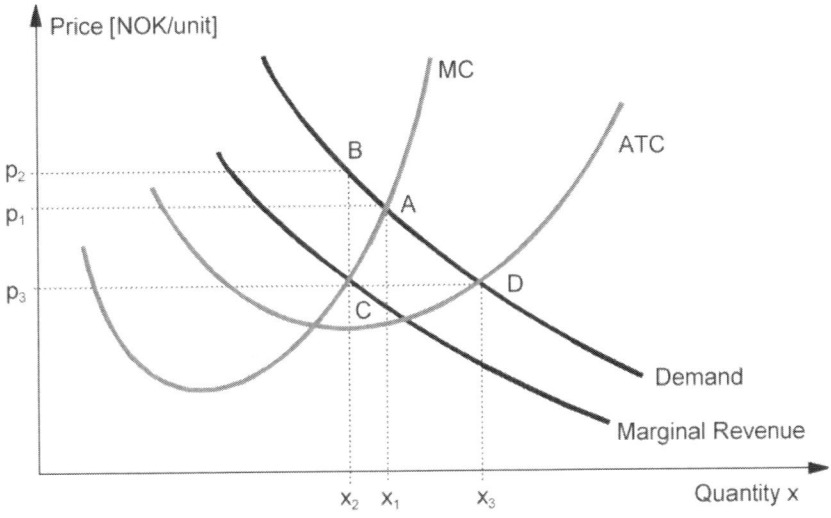

Figure 2.8 Monopolistic competition.

2.8.2 Oligopoly

If there is more than one supplier on the market, there is still a possibility that some of them can affect the market price, although not to the same extent as a monopolist. We call this situation an *oligopoly*.

Different models have been developed to describe the oligopoly. One of the most frequently used is the Cournot model[5], which is fairly simple and based on assumptions that are regarded as being realistic.

The Cournot model is described by a simple example from [1], which is illustrated in Figure 2.9.

---

[5]Based on the French economist Augustin Cournot, which in 1838 described this form of oligopolistic competition. The example described here is the same as the one Cournot used in his original work.

Chapter 2: Basic microeconomics

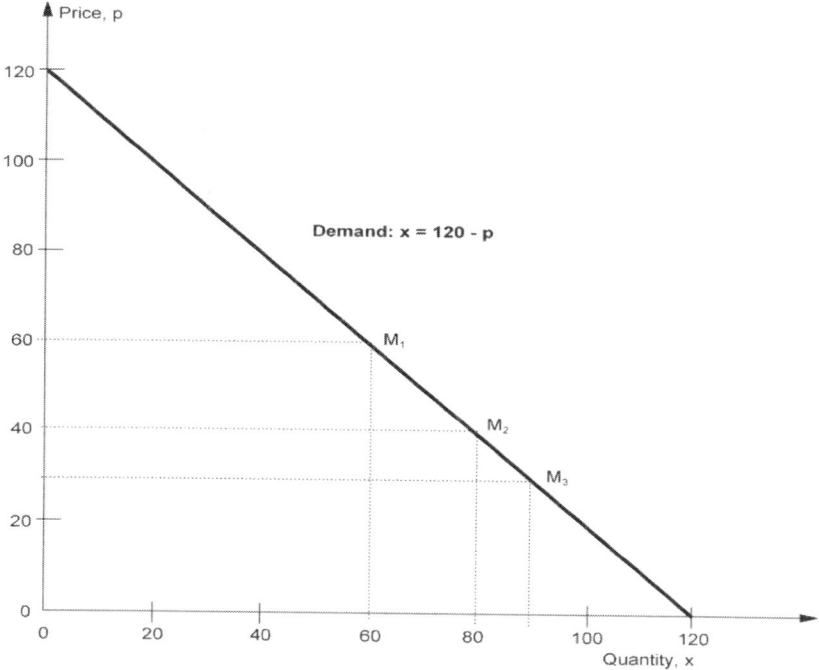

Figure 2.9 Falling demand curve and solutions for a Cournot-oligopoly.

We assume zero cost[6] and a linear demand curve:

$$p = 120 - x \quad (2.25)$$

As long as there is no production cost, an efficient market will give a price equal to zero and a production quantity of 120, which is the maximum welfare solution.

If production is controlled by one or a few companies, it is possible to reduce output and thereby raise the market price in order to earn some profit. The simplest example is still the monopolist, which could be a cartel or one single company, controlling the whole production. Maximizing the revenue (which in this zero cost case equals the profit) for the monopolist gives:

$$\pi = xp = x(120 - x) \quad (2.26)$$

---

[6] The original example was a natural spring supplying water.

Imperfect competition (market power)

It is easy to find the first order condition for maximum revenue, and the solution is:

Quantity: $x = 60$
Price: $p = 60$
Revenue: $\pi = 3600$

This solution is marked as M_1 in Figure 2.9. From the figure we can also see that the welfare loss (compared with an efficient solution) is 1800.

Next we assume there are two competing companies, both with a profit maximization objective. The revenue for the two producers will be:

$$\pi_1 = x_1(120 - x_1 - x_2)$$
$$\pi_2 = x_2(120 - x_1 - x_2) \tag{2.27}$$

The Cournot-model for this oligopoly is based on two assumptions:

- The producers know the market demand curve.
- Each of the producers decides its own output as if output from the other producer (or producers) is given.

We can then find the first order condition for maximum revenue (and profit) for the two companies.

$$\partial \pi_1 / \partial x_1 = 120 - 2x_1 - x_2 = 0$$
$$\partial \pi_2 / \partial x_2 = 120 - x_1 - 2x_2 = 0 \tag{2.28}$$

This set of equations can be solved with respect to the two production volumes.

Production volumes: $x_1 = x_2 = 40$
Price: $p = 40$
Revenue (profit): $\pi_1 = \pi_2 = 1600$

The total volume is here 80, and the welfare loss (compared to an efficient solution) amounts to 800. The solution is marked as M_2 in Figure 2.9.

If we increase the number of competing companies, we can find a general solution for how one of them will adapt. For unit number 1, the solution will be (in principle the same as Equation (2.28),

$$\partial \pi_1 / \partial x_1 = 120 - 2x_1 - x_{rest} = 0 \tag{2.29}$$

where x_{rest} is the volume from all the other producers. As long as all the companies have the same market share and adapt in the same way, this volume will be:

$$x_{rest} = (n-1) x_1 \qquad (2.30)$$

This leads to the following equation for an arbitrary number of units.

$$120 - 2x_1 - (n-1) x_1 = 0 \qquad (2.31)$$

Table 2.1 Cournot competition with increasing number of producers.

No. of units	1	2	3	4	10	∞
Total quantity	60	80	90	96	109	120
Price	60	40	30	24	11	0
Social welfare loss	1800	800	450	288	60	0
Profit for producers	3600	3200	2700	2304	1199	0

Table 2.1 shows the results with an increasing number of companies competing under these assumptions. As the number of companies increases, the total quantity delivered increases, while the price, the social welfare loss and the profit for producers decrease. With an infinite number of units, we end up with an efficient solution, i.e. a perfectly competitive market.

This simple model is based on a zero production cost, which is normally not the case. The general equation for Cournot-model is:

$$\partial \pi_i / \partial x_i = p + x_i (\partial p / \partial x_i) - MC(x_i) = 0 \qquad (2.32)$$

2.9 Externalities

The term *external cost* or *externality*, refers to a part of the production cost which is not paid by the firm producing a commodity, but by external persons or companies being affected. The classical example of externalities is the costs of pollution. This is a cost to society, but not necessarily to the firm causing the pollution.

Externalities cause inefficiency even in a perfectly competitive market if it is not compensated for. One way to neutralize externalities is through taxation. This idea was introduced by the English economist Pigou as early as the 1920s. It is of course a problem to assess the consequences of pollution or other externalities correctly in economic terms. However, if we assume we have a correct assessment, we can put a tax on the producers as shown in Figure 2.10.

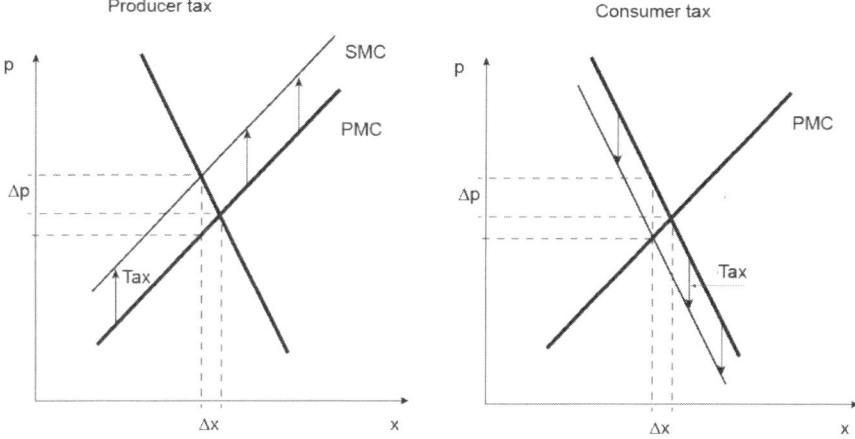

Figure 2.10 Taxation reflecting external cost.

The tax will increase the marginal cost of the producers from PMC (Private Marginal Cost) to SMC (Social Marginal Cost), and thereby cause a shift in the supply curve as indicated in the figure. The price will increase and the quantity decreases accordingly.

Notice that the final price impact - the price seen by the consumer - of a tax (Δp) is the same for a producer tax as for a consumer tax. The price impact depends on the elasticity, not on who is primarily carrying the tax. Notice also the quantity impact (Δx) is the same for the two cases.

If the external cost is equal for all producers (per unit produced), the tax can either be passed on the producers or the consumers. It is easy to see that the effect will be the same. However, this is generally not the case. Some producers cause more pollution than others. Consequently, the general rule is that environmental taxes should be placed on the producers. The so-called "Polluter Pays Principle" is a well-known principle in this context. The firm causing the pollution should be taxed according to the damage.

These Pigouvian taxes have two effects. Firstly, the taxes cause an efficient allocation of resources as indicated in Figure 2.10. The price after taxation equals the social marginal cost. Secondly, the taxes give public income. In general, taxation is a problem because it normally causes economic inefficiency. However, if a Pegouvian tax is introduced, other taxes can be reduced accordingly and the corresponding inefficiency is also reduced. Due to this, it is claimed that environmental taxation gives double dividend. [2]

Another possibility to avoid or limit external effects is to use restrictions. Other instruments can also be used. Chapter 10 describes different tools at hand in the power sector.

2.10 References

[1] Walter Nicolsen: "Microeconomic Theory", The Dryden Press, 1992.

[2] Agnar Sandmo: "The Public Economics of the Environment", Oxford University Press, 2000.

3 Electricity consumption

3.1 General

The electricity consuming sector is a part of the power system. Power system analysis has traditionally concentrated on the supply side, but it is important to include the demand side as well. The consumption is variable over time with a characteristic pattern for different consumer categories. It is affected by price and other factors.

3.1.1 Characterization

The continuity and the variability of electricity consumption, opens different alternatives for characterization of the load. (Load in this case means not only consumption but generally a flow of electricity.) Which parameters do we describe it by and which units is it measured by? The most common alternatives are listed in the table below.

Table 3.1 Characterization of electricity consumption.

Characteristic	Unit (example)
Instant load	kW
Energy consumption during a certain time	$kWh/year$
Maximum load	kW_{max}
Energy and max load (load factor)	$kWh/year$ and kW_{max}
Load-duration curve	
Load curve (chronological) with different time resolution	

The choice of characterization depends on the purpose or the type of analysis we use it for. Instant load can be useful in an analysis of the spot market. In some cases, for instance analysis or long-term trends, electricity consumption per year can be more relevant. The maximum load is interesting for the design of supply side components such as transformers and lines.

For a more thorough analysis, the energy consumption and maximum load are needed, in some cases a load-duration curve. A typical load-duration curve is shown in *Figure* 3.1. Duration can be given in hours or as percentage of total time. The area under the load duration curve is the total energy consumption. We define the *load utilisation time* as:

$$T_b = \frac{E}{P_{max}} \qquad (3.1)$$

and the *load factor* as:

$$A = \frac{T_b}{8760} \qquad (3.2)$$

Figure *3.1* Load-duration curve for a period of one year.

The load-duration curve gives only limited information about the load variability. In many cases a (chronological) load curve is needed. Load curves for different time spans and different time resolutions are shown later.

3.1.2 Categorization

Electricity consumption can be divided into different categories based on different criteria as indicated in Table 3.2.

Table 3.2 Alternatives for categorization of electricity consumption.

By sector:	1. Household	2. Service	3. Industry	Etc.
By use:	1. Heat	2. Light	3. Appliances	Etc.
By firmness:	1. Firm power	2. Curtailable	Etc	
By size:	1. Large	2. Medium	3. Small	
Etc				

In many statistical surveys consumption is divided by sector, but other types of categorization can be interesting in many cases.

3.2 Demand drivers

It is possible to identify different driving forces, on a macro as well as a micro level, underlying the development of electricity demand.

3.2.1 Income elasticity

On a macro level, the general economic activity in terms of Gross Domestic Product (GDP), seems to be the major driving force behind the development of electricity consumption.

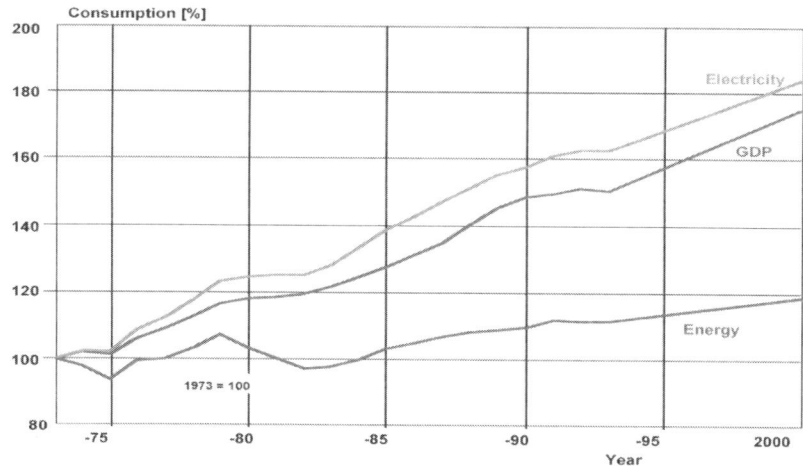

Figure 3.2 Development of energy and electricity consumption vs. GDP for 12 EU member states[1]. Source UNIPEDE.

Figure 3.2 gives a clear indication that the electricity consumption follows the GDP very closely. The growth is slightly higher than the GDP growth, indicating an income elasticity slightly higher than one (we assume the income is proportional to GDP).

According to the figure, the energy (here the Total Primary Energy, TPE) consumption is not so closely linked to GDP. GDP can increase at the same time as the energy consumption decreases. We should remember, however, that in this particular period, the world oil market was highly volatile. On two different occasions OPEC embargoes lead to dramatic increases in the oil prices. This had an impact on energy consumption and had also consequences for the total output of the economy, which can be seen from the figure. The electricity prices were not so much affected and electricity consumption continued to follow the general activity trend.

As already mentioned, the income elasticity for electricity has been slightly higher than one.

This has been typical for industrial nations. As we move into a later stage (a post-industrial stage) the elasticity falls below one.

For countries in an early stage of development income elasticity is normally considerably higher. Table 3.3 shows some examples of energy and electricity

---

[1] The 12 "traditional" EU member states: Austria, Belgium, Denmark, France, Germany, Greece, Ireland, Italy, Luxembourg, Netherlands, Portugal and Spain

intensities in some industrial countries, a former communist country and two developing countries.

Table 3.3 Energy intensities in tones oil equivalents (Toe) per USD 1000 GDP and electricity intensities in kWh per USD GDP[2].

	Total energy Toe/USD1000 PPP	Electricity kWh/USD PPP
OECD	0.27	0.43
Norway	0.23	1.12
Sweden	0.33	0.91
USA	0.34	0.75
Germany	0.25	0.39
India	0.37	0.32
China	0.31	0.29

Energy and electricity consumption per capita is much lower in the developing countries, India and China, than in the industrial ones. But the energy intensity is higher. The electricity intensity on the other hand is higher in the industrial countries. So if the developing countries are going to follow a growth pattern like the industrial countries, we should expect a growth in energy consumption that is lower than the general economic growth and a growth in electricity consumption that is higher than the growth in GDP. This is in line with the developments we have seen during the last few years. See [1] and [2].

We notice that the energy intensity is higher in the Russia than in the western industrial economies. The electricity intensity is extremely high. That indicates economic inefficiencies that will probably vanish in the future.

The USA has also a relatively high energy intensity, but not to the same extent as Russia. Low prices can be plausible reason.

We also notice Norway's energy intensity is at about the OECD average, but the electricity intensity is extremely high. That reflects both household and industrial consumption of electricity.

---

[2] PPP is short for Purchasing Power Parity, which means that the value of the GDP is referred to the country's price level. That is a better way to indicate the output of the economy than just using the exchange rate related to USD.

3.2.2 Price elasticity

As for most other commodities, electricity consumption depends on price. Several studies have been done to investigate the price elasticity for electricity. The conclusion from these studies is that the elasticity is less than 0, as one should expect, but the estimated numbers are quite different. There seems to be different price elasticity in different parts of the world and in different sectors of the economy. There is also a difference between long-term and short-term elasticity, but generally the results from the studies are quite divergent. Some of the divergent results are caused by the fact that electricity consumption depends on several factors in addition to price. That makes it difficult to estimate the impact of price alone.

We broadly distinguish between two alternative procedures for the estimation of price elasticities: One is *time series analysis*, where we follow a group of consumers through a certain time and observe how consumption changes with price. The other is *cross section analysis*, where we compare different groups of consumers at one time and observe to what extent price differences cause differences in consumption. We can also combine these two procedures.

There are different problems connected to the two procedures. The major problem is to exclude the impact of other factors and find the effect of prices alone. A time series analysis is affected by changes in income, technology and other factors over time. Some of these factors can be taken into account, but not fully compensated for. In addition we often have the problem that prices can be relatively stable over time. (At least that was a problem before deregulation.) That makes a precise estimation of price effects difficult. Cross section analysis faces similar problems. It is difficult to find comparable consumer groups where all factors are equal except price.

We will present the results from a few studies here. We will not go into detail concerning the analysis, but simply quote the results. We refer to [3] and [4] for a more comprehensive overview.

In a study from Canada (see Table 3.4) it is clearly demonstrated that short-term elasticities are lower (in absolute value) than the long-term ones. We also notice that households are more price sensitive than the industry and service sectors.

Table 3.4 Results from Canadian studies on price elasticity.

Study	Sector	Price elasticity	
		Short term	Long term
National Energy Board, Canada 1989	Household	-0.16	-0.73
	Service	-0.13	-0.46
	Industry	-0.11	-0.45
Energy, Mines & Resources, Canada 1990	Household	-0.14	-1.10
	Service	-0.06	-0.36
	Industry	-0.09	-0.49

A summary of 8 different studies from the USA is presented in Table 3.5. We notice a certain similarity with the results from Canada, but the results are generally more divergent.

Table 3.5. Price elasticities for the USA.

Sector	Price elasticity	
	Short term	Long term
Household	-0.20	-1.00 to -1.80
Industry	-------	-0.40 to -1.56

The same is the case with studies of different OECD countries presented in Table 3.6. In this case one has looked at different countries and different sectors, but there is not always distinction between long term and short term.

Table 3.6 Estimated price elasticities.

Study	Sector	Country	Price elasticity
V.B. Hall, 1986 Long term OECD countries 1969 - 79	Industry	USA	-0.14
		Germany	-0.30
		Japan	-0.10
Party, Nappy and Tagvai G -7 1960 - 87	Household	Great Britain	-0.17
		France	-0.66
		USA	-1.03
	Industry and service	Great Britain	-1.03
		France	-1.63
		Germany	-0.75
		USA	-0.55
Longva, Olsen and Strøm, 1988	Household	Norway	-0.53
	Industry	Norway	-0.65

All these examples demonstrate that electricity consumption is price dependent, but the level of the price elasticity is highly uncertain. There is more elasticity in the long run than in the short run. It is difficult to draw a certain conclusion concerning the differences between sectors.

Some studies have also included *cross price elasticities* for oil, gas and electricity. The Canadian study (NEB) found that the cross elasticity for electricity with respect to the oil price was 1.57 for households. That indicates that a price increase of 1% on oil will lead to 1.57% increase in electricity consumption.

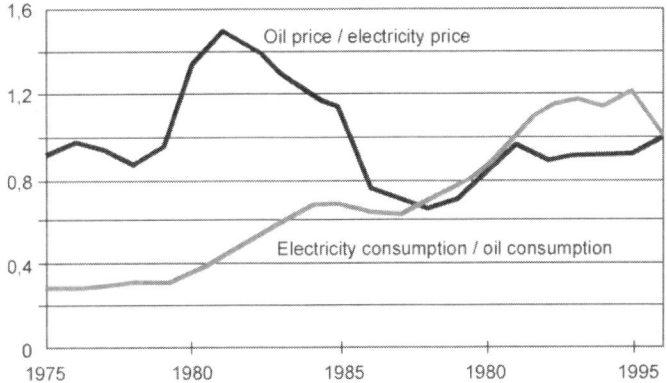

Figure 3.3 Development of relative price and quantity for oil and electricity in Norway. Index, 1996 = 1.

Figure 3.3 shows how the relative price for oil/electricity and the relative consumption for electricity/oil developed between 1975 and 1996. It is obvious that high oil prices lead to increased electricity consumption.

With respect to prices or tariffs with variation over time, it can be relevant to define a *cross time elasticity* for electricity. Cross time elasticity indicates that a price change at one point in time can affect consumption at another time. We return to this matter in Chapter 5.

3.3 Load research

Load research comprises different activities with the purpose of gathering information and analyse the demand side of the power system. We normally divide it into three main areas: 1) load measurements to investigate end-use load patterns, 2) investigation of factors (other than price) affecting end-use and 3) load aggregation. In this section we describe some results from load research activities in Norway in the years 1980 to 1992.

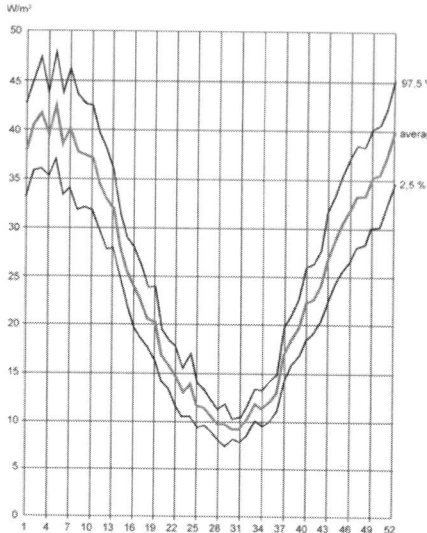

Figure 3.4 Yearly load profile. Household.

Figure 3.5 Yearly load profile. Office buildings.

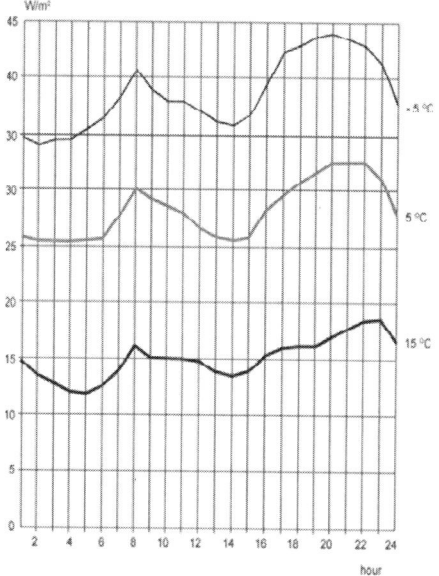

Figure 3.6 Day/night profile. Households.

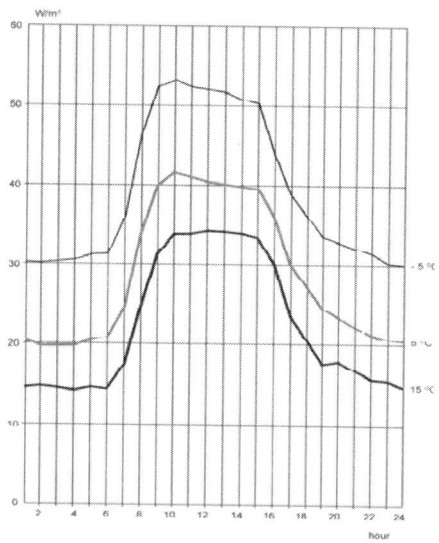

Figure 3.7 Day/night profile. Office buildings.

3.3.1 Load measurements

Various end-users have different load patterns. Each consumer has a characteristic variability over a year, over a week and over a typical day/night. Consumers can be grouped into classes with similar load patterns. Correlation between different consumers and different classes of consumers is interesting with respect to load aggregation. We will come back to this later.

Figure 3.4 and Figure 3.5 show yearly load (W/m^2) variation for single family houses and office buildings. Norway has a typical winter peak due to electrical heating. Notice that the annual variation and the average level are similar, but office buildings have a larger variation between different buildings.

Day/night profiles are shown in Figure 3.6 and Figure 3.7. These profiles are fundamentally different for the two categories. Households have a typical short duration morning peak at about 8 and an afternoon/evening peak which is higher and longer. Offices have high load during the working hours with a peak in the morning at about 10.

Notice that the curves in Figure 3.6 and 3.7 indicate temperature dependency whereas the curves in Figure 3.4 and 3.5 indicate random variation in the load. We notice that temperature dependency is higher for households than for offices, especially for high temperatures. The difference between 5 °C and 15 °C is substantial for single family houses, but smaller for offices.

3.3.2 Factors influencing the load level

As already indicated in Figures 3.4 to 3.7, the temperature is an important factor for the load. In Norway, where we have a dominant share of electric space heating (approximately 70%), the load will increase with decreasing temperature. In other countries, where air conditioning is more important, the load will increase with rising temperature and the summer peak is higher than the winter peak.

There are two different ways in which the temperature affects the load. One is the peak load which is depends the minimum temperature (or the maximum in summer peak systems). The other is accumulated need for heating in terms of kWh per year, which depends on the so-called "degree-days". In Norway degree-days are defined as shown in Figure 3.8. The figure indicates that the heating season starts in the autumn when the temperature falls below 11 °C and ends in the spring with temperatures above 9 °C. This "unsymmetrical" heating requirement is due to contribution from the sun which is stronger in the spring.

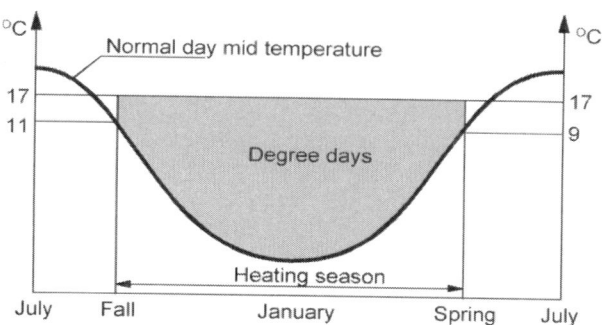

Figure 3.8. Degree-days.

In addition to temperature, electricity consumption depends to some extent on wind, and cloudiness, but that effect is very limited.

Measured temperature dependency for different building categories is shown in Table 3.7

Table 3.7 Temperature dependency for different building categories.

Category	Temperature dependency	
	Peak load $(W/m^2\ °C)$	Energy consumption $(Wh/m^2\ degree\text{-}days)$
Block of flats	0.96	18.6
Single family house	1.21	25.2
Office building	1,44	29.8
Hospital	0.99	20.0

3.3.3 Load aggregation

The load from a set of consuming entities will accumulate as more entities are added. As one moves from a local feeder to higher grid levels the load will increase, and the load shape will change. If we use instant load or energy per year as load characterization, the accumulated or aggregated load can be found simply by adding the individual loads. But if the load is described by a load-duration curve or by maximum load and corresponding load factor, the problem of finding the resulting load, becomes more complex.

The problem is illustrated by the Figure 3.9. and Figure 3.10. We have set of individual loads that add up at a branching point as shown in Figure 3.9. Load variation on the individual and the aggregated level is shown on the next figure.

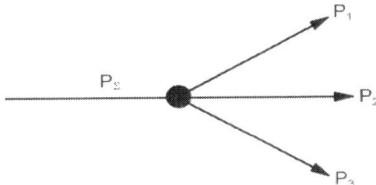

Figure 3.9 Grid branching point.

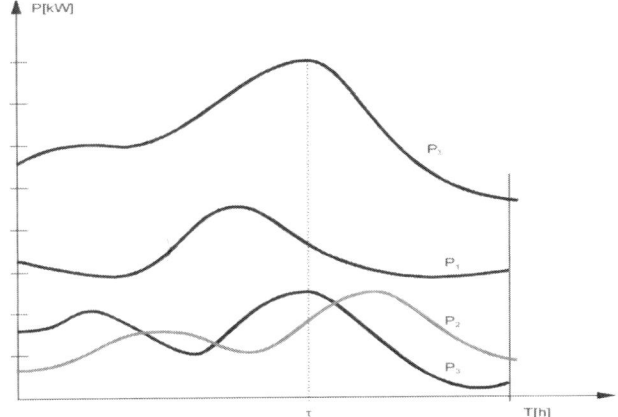

Figure 3.10 Individual and aggregated load curves.

It is easy to see that the aggregated maximum load is less than (or at the most equal to) the sum of the individual maximum loads because the individual maxima are not coincident. We define a load coincidence factor (or simply a coincidence factor) for this relationship.

The load for the individual consumers is a function of time as is the resulting load:

$P_1(t)$ = load consumer 1

$P_2(t)$ = load consumer 2

$P_3(t)$ = load consumer 3

$P_\Sigma(t)$ = resulting load consumer 1 plus 2 plus 3

The power balance at each instant of time requires that (we disregard losses):

$$P_\Sigma(t) = P_1(t) + P_2(t) + P_3(t) \qquad (3.3)$$

From Figure 3.10 we can see that:

$$P_{\Sigma max} = P_{\Sigma}(\tau) = P_1(\tau) + P_2(\tau) + P_3(\tau) \qquad (3.4)$$

and:

$$P_{\Sigma max} \leq P_{1max} + P_{2max} + P_{3max} \qquad (3.5)$$

We can then find a set of coincidence factors, s_1, s_2 and s_3, representing each individual maximum's share of the aggregated maximum:

$$P_{\Sigma max} = s_1 P_{1max} + s_2 P_{2max} + s_3 P_{3max} \qquad (3.6)$$

If we combine Equation (3.6) with Equation (3.4), we see that:

$$s_1 = \frac{P_1(\tau)}{P_{1max}}$$

$$s_2 = \frac{P_2(\tau)}{P_{2max}} \qquad (3.7)$$

$$s_3 = \frac{P_3(\tau)}{P_{3max}}$$

Thus the coincidence factors (based on this deterministic description) are equal to the individual load at peaking hour (of the aggregated load) divided by the individual maximum load.

But the load is stochastic to some extent. It is affected by stochastic incidents, relatively more on an individual level than on an aggregated one. Figure 3.11 shows peaking power per customer for an increasing number of aggregated households. The peaking power per customer decreases which means that the load factor increases with increasing load aggregation. The figure also shows that the stochastic variation declines as more customers are added.

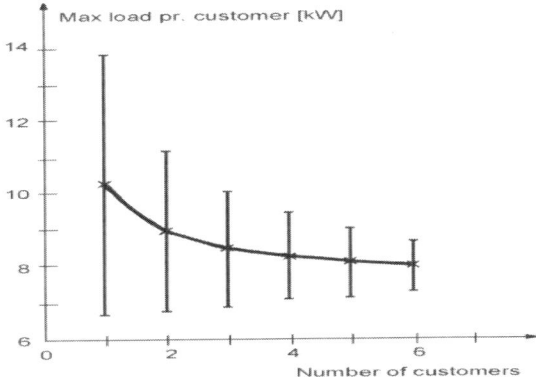

Figure 3.11 Maximum (aggregated) load per customer with increasing number of customers. Variance is indicated with vertical bars. Norwegian household customers.

Based on this stochastic load description, it is reasonable to define a coincidence factor as:

$$s = \frac{P_{\Sigma max}}{\sum_{i=1}^{n} P_{i,max}} \qquad (3.8)$$

This coincidence factor is not tied to individual loads as the ones defined by Equation (3.7). It is a general coincidence factor for a certain category of consumers.

Load aggregation, or more precisely estimation of aggregated load characteristics (in particular the peak load) is a necessary part of a planning process. There are different procedures for dealing with load aggregation. We will describe three different types here:

- Use of constant coincidence factors
- Use of Velander's formula
- Procedure based on statistical analysis (USELOAD)

Chapter 3: Electricity consumption

Coincidence factors

When we are using the coincidence factors as they were defined above (Equation (3.8)) for load aggregation, we are making the assumption that the coincidence factor at a given grid level can be treated as a constant. The coincidence factors are estimated by data from load measurements at the different grid levels. Based on the estimated coincidence factors, the peak load at different grid levels can be found through the following procedure:

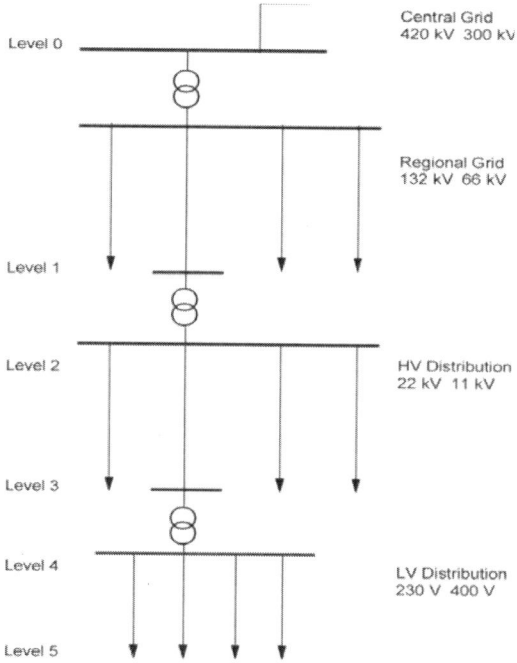

Figure 3.12 Different grid levels.

Figure 3.12 shows a distribution grid with levels marked from 0 to 5[3].

---

[3] Designations are according to definitions used by the Norwegian regulator. Level 5 represents consumers connected to the low voltage distribution grid. Level 3 represent high voltage distribution. See also table 3.12.

$$P_{4,1,max} = s_5 \sum_{i=1}^{n_{5,1}} P_{5,1,i,max}$$

$$P_{4,2,max} = s_5 \sum_{i=1}^{n_{5,2}} P_{5,2,i,max} \qquad (3.9)$$

$$P_{4,3,max} = s_5 \sum_{i=1}^{n_{5,3}} P_{5,3,i,max}$$

In the same way we find the maximum load at level 3 as:

$$P_{3,max} = s_4 \sum_{i=1}^{n_{4,1}} P_{4,i,max} \qquad (3.10)$$

And combining the equations above we find that:

$$P_{3,max} = s_5 s_4 \left[\sum_{i=1}^{n_{5,1}} P_{5,1,i,max} + \sum_{i=1}^{n_{5,2}} P_{5,2,i,max} + \sum_{i=1}^{n_{5,3}} P_{5,3,i,max} \right] \qquad (3.11)$$

This means we find the resulting maximum at level 3 by adding all maximum loads at level 5 and multiplying by s_5 and s_4. And generally we find a resulting coincidence factor from level i to level 1 as the product:

$$s_{res} = s_1 \, s_2 \, s_3 \, ... s_i \qquad (3.12)$$

For a transmission/distribution grid with levels as shown in Figure 3.12, the coincidence factors can be arranged in a matrix as shown in Table 3.8.

Table 3.9 gives typical numerical values for households in Norway.

Again it should be emphasized that we assume constant coincidence factor at each grid level. This is in fact a critical assumption for this procedure. It is acceptable as long as we have a uniform load, for instance only household consumption in the whole area. But if we have a mix of different categories, the assumption is not acceptable.

Table 3.8. Matrix of coincidence factors for different grid levels.

Level	0	1	2	3	4	5
Central grid (0)	1.0	s_1	$s_1 s_2$	$s_1 s_2 s_3$	$s_1 s_2 ... s_4$	$s_1 s_2 ... s_5$
Regional grid (1)		1.0	s_2	$s_2 s_3$	$s_2 s_3 s_4$	$s_2 s_3 ... s_5$
HV transformer (2)			1.0	s_3	$s_3 s_4$	$s_3 s_4 s_5$
HV distribution grid (3)				1.0	s_4	$s_4 s_5$
LV transformer (4)					1.0	s_5
LV distribution (5)						1.0

Table 3.9 Matrix of coincidence factor for different grid levels. Numerical values.

Level	0	1	2	3	4	5
Central grid (0)	1.0	0,96	0.94	0.85	0.83	0.58
Regional grid (1)		1.0	0.98	0.88	0.86	0.61
HV transformer (2)			1.0	0.90	0.88	0.62
HV distribution grid (3)				1.0	0.98	0.68
LV transformer (4)					1.0	0.70
LV distribution (5)						1.0

Velander's formula

The coincidence factors are peak-kW-to-peak-kW conversion factors. In some cases it can be more convenient to use a kWh-to-peak-kW conversion factor. That is relevant if the total kWh consumption of the entities is known. It is then possible by use of this kWh-to-peak-kW conversion factor (sometimes called a C factor) to make an estimate of the peak load. This C factor can be known from experience or more or less systematic load research.

A special formula has – and is still to some extent – used for this conversion factor. It is called Velander's[4] formula:

$$P_{max} = k_1 E + k_2 \sqrt{E} \qquad (3.13)$$

E is the total energy consumption and k1 and k2 are constants that have to be estimated. Velander's formula is basically an empirical formula, but it can be also be derived from mathematical statistics.

Constants k1 and k2 that have been estimated from load measurements in Norway, are shown in Table 3.10.

Table 3.10 Estimated Velander constants for households in Norway.

Region	k_1	k_2
Eastern	0.00022	0.019
Western	0.00022	0.015
Central Norway	0.00021	0.024

As the number of customers and the energy E increase, the relative contribution of the square root term in Velander's formula will drop and we will approach a situation where P_{max} is proportional to E. That is indicated in Figure 3.11. When we reach about 30 customers, the contribution of each new customer (to the aggregated load) is about 7 kW whereas each individual customer has a maximum load of 8 -14 kW.

Utilization time for the aggregated load:

$$t_b = 1/k_1 \qquad (3.14)$$

With the data above, the utilisation time is 4545 hours corresponding to load factor of 0.52.

For a household with 30 000 kWh/year (typical Norwegian consumption), Velander's formula gives P_{max} of 10 kW for that consumer alone, but as contribution to the aggregated P_{max} it is 6.6 kW. This corresponds to a coincidence factor of 0.62 which can be compared with the numerical values in Table 3.9.

---

[4]It is named after the Swedish professor Velander. In other parts of the world similar formulas are used, but under other names.

Statistical analysis (USELOAD)

There are some drawbacks in the two methods described above. They are both based on simplifications and the serious disadvantage that they are unable to handle a mix of diverse load categories. A method based on statistical analysis where statistical data from load measurements are made use of, represents an improvement in this respect. The program USELOAD, developed by SINTEF Energy Research, [5] and [7] is an example of this kind of tools.

Assume the power consumption for one end-user at a given time can be described as a random variable with a normal distribution[5] as indicated in Figure 3.13.

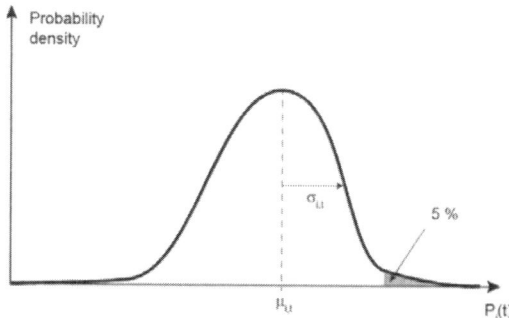

Figure 3.13. Normal distribution with an indicated 95% confidence limit

$P_i(t)$ is the consumption of end-user i at time t. $\mu_{i,t}$ and $\sigma_{i,t}$ are expected value and standard deviation for $P_i(t)$.

In order to define the peak consumption for this description of the load as a random variable, we have to introduce a probability that that the load will not exceed that peak level. We use the following formula:

$$P_{i,\max} = \mu_{i,\tau} + k\sigma_{i,\tau} \qquad (3.15)$$

where $P_{i,\max}$ is the maximum which is expected to occur at the time τ, $\mu_{i,\tau}$ is the corresponding expected value and $\sigma_{i,\tau}$ is the standard deviation at that time. The factor k is decided by the probability we will tie to our definition of peak load. If we for instance decide that probability of exceeding the peak is 5%, the factor k

---

[5] One could assume other distributions instead the normal distribution, and the procedure would largely be the same. But normal distribution is convenient for practical reasons.

will be 1.65 (provided normal distribution). A higher factor corresponds to lower probability.

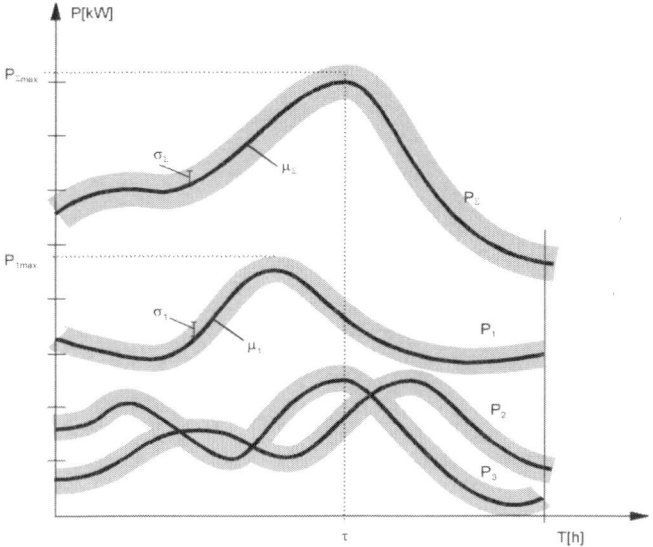

Figure 3.14 Individual and aggregated load curves with stochastic variation indicated.

As long as all the individual loads are normally distributed, the aggregated load at a given time will also be normally distributed. If we look on the peaking hour (τ), that will also be the case. The total (aggregated) load $P_\Sigma(\tau)$, will have a normal distribution:

$$N(P_\Sigma(\tau); \mu_{\Sigma,\tau}, \sigma_{\Sigma,\tau}) \qquad (3.16)$$

where we can find the expectation $\mu_{\Sigma,\tau}$ and the standard deviation $\sigma_{\Sigma,\tau}$ by the formula:

$$\mu_{\Sigma,\tau} = \sum_{i=1}^{n} \mu_{i,\tau} \qquad (3.17)$$

and

$$\sigma_{\Sigma,\tau}^2 = \sum_{i=1}^{n} \sum_{j=1}^{n} \rho_{i,j,\tau} \sigma_{i,\tau} \sigma_{j,\tau} \qquad (3.18)$$

where $\rho_{i,j,\tau}$ is the coefficient of correlation between load i and load j at time τ. This correlation coefficient is 1 for all i = j (every load is perfectly correlated with itself) and for all i ≠ j it is a number between -1 and +1. Zero means there is no correlation between the two variables.

Once the expectation $\mu_{\Sigma,\tau}$ and the standard deviation $\sigma_{\Sigma,\tau}$ are found, we can find the aggregated peak load by a formula equivalent to Equation (3.15):

$$P_{\Sigma,\max} = \mu_{\Sigma,\tau} + k\sigma_{\Sigma,\tau} \qquad (3.19)$$

All the input variables to Equation (3.18) and (3.17) i.e. expected values, standard deviations and correlation coefficients can be found by load measurements.

Example with two loads

We use a simple example to illustrate the procedure. We regard two loads and assume stochastic variation. The load at a given time is characterized by the expected value and the variance for each one and a correlation coefficient expressing the correlation between the two.

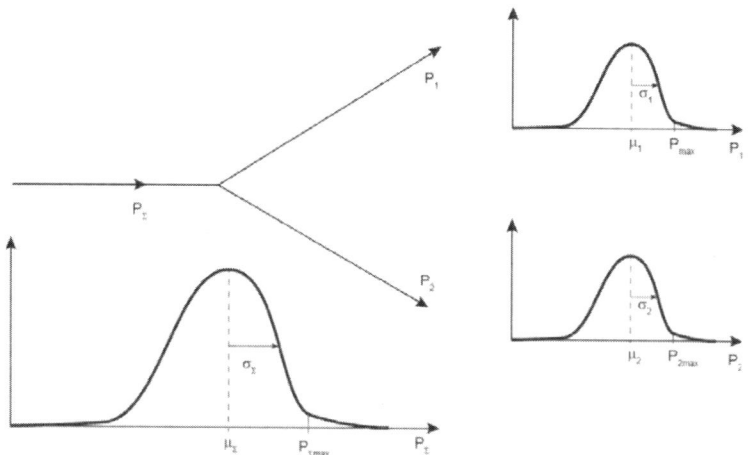

Figure 3.15. Two loads added to an aggregated load

The maximum for the aggregated load is then given by the formula:

$$P_{\Sigma,\max} = \mu_{1,\tau} + \mu_{2,\tau} + k\sigma_{\Sigma,\tau} \qquad (3.20)$$

where:

$$\sigma^2_{\Sigma,\tau} = \sigma^2_{1,\tau} + \sigma^2_{2,\tau} + \rho_{1,2,\tau}\sigma_{1,\tau}\sigma_{2,\tau} \qquad (3.21)$$

where $\rho_{i,j,\tau}$ is the coefficient of correlation between load i and load j at time τ.

The parameters are estimated on the basis of observations (measurements). Assume that we have recorded the load at a given hour (the peaking hour) during a number of comparable days. We make *n* observations and we are using the formulas normally used to estimate the statistical parameters: The mean for each of the two samples represents the expectation:

$$\mu_1 \approx \overline{P_1} = \frac{1}{n}\sum_{i=1}^{n} P_{1i} \qquad (3.22)$$

$$\mu_2 \approx \overline{P_2} = \frac{1}{n}\sum_{i=1}^{n} P_{2i} \qquad (3.23)$$

The variance represents the standard deviation for the observations and can be expressed as:

$$\sigma_1 \approx v_1 = \sqrt{\frac{\sum_{i=1}^{n}(P_{1i} - \overline{P_1})^2}{n-1}} \qquad (3.24)$$

$$\sigma_2 \approx v_2 = \sqrt{\frac{\sum_{i=1}^{n}(P_{2i} - \overline{P_2})^2}{n-1}} \qquad (3.25)$$

The correlation coefficient is represented by the sample correlation coefficient:

$$\rho_{1,2} \approx r_{1,2} = \frac{\sum_{i=1}^{n}(P_{1i} - \overline{P_1})(P_{2i} - \overline{P_2})}{nv_1 v_2} \quad (3.26)$$

As indicated in the equations, the formulae represent approximations. The expected level of the resulting load can then be estimated as:

$$\mu_\Sigma = \mu_1 + \mu_2 \quad (3.27)$$

and the estimated standard deviation:

$$\sigma_\Sigma^2 = \sigma_1^2 + \sigma_2^2 + \rho_{1,2}\sigma_1\sigma_2 \quad (3.28)$$

A simple numerical example is shown in Table 3.11. We have recorded the load for two consumers at a given hour (the peak hour) for 6 comparable days. Table 3.11 shows the recorded loads, the mean values, the variances and the estimated correlation coefficient based on the formulas above.

Table 3.11 Recorded maximum load on 6 different days for two customers.

Day	1	2	3	4	5	6		Expectation	Standard Deviation
Consumer 1	1	2	3	3	2	1	(kW)	2,000	0,89442719
Consumer 2	10	20	30	30	20	10	(kW)	20,0000	8,94427191
Sum		11	22	33	33	22	11	(kW)	22,0000	

The correlation coefficient is 0.833

We can now find the resulting variance from Equation (3.28):

$$\sigma_\Sigma = \sqrt{(0,8944^2 + 8,944^2) + 0,833 \cdot 0,8944 \cdot 8,944} = \sqrt{87,46} = 9,35 \ (kW)$$

If we define the peak load as the load tied to the 95% probability level, we get the following formula for the calculation of the resulting peak load:

$$P_{\Sigma Max} = 1,645 \cdot \sigma_\Sigma + \mu_\Sigma = 1,645 \cdot 9,35 + 22,0 = \underline{37.38\ (kW)}$$

We see that this calculation gives a higher estimate of the resulting peak load than a direct sum. (38 compared to 33 kW.)

In this simple example we are only making use of a few observations. That means the formulas represent rough approximations.[6] In a more realistic example, that will not be the case.

Example of peak load calculation by use of USELOAD

The pc tool USELOAD makes use of statistical data describing the load. These data are collected through a long period of time and parameters are estimated based on that. One advantage of USELOAD is that it can handle an arbitrary mix of different consumer categories.

Input and output

Data provided by the user:

- Data on yearly energy consumption for the different consumer categories
- Local climate information.

Data from the data base:

- Load curves for different consumer categories
- Energy intensities and temperature dependencies.
- Standard deviation and correlation coefficients

Output can be presented in as graphs (Figure 3.16) and as Tables (3.12)

---

[6] There are statistical methods available for cases where we have a limited number of observations. Such methods could in principle be used, but in practice methods based on the described approximation has proved to be satisfactory.

Chapter 3: Electricity consumption

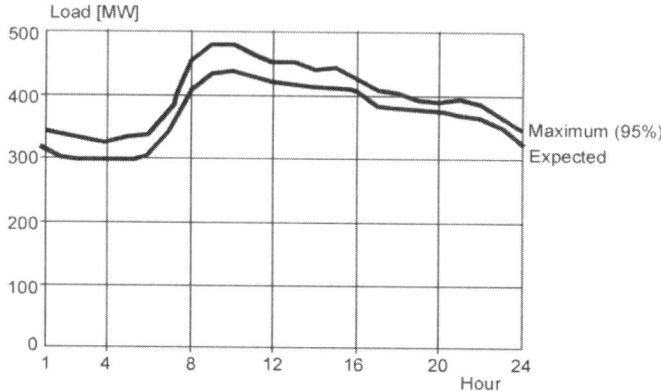

Figure 3.16 Load curve.

Table 3.12 shows that residential customers (H4 tariff) service sector customers (T32) have similar contributions to the peak load in this network (44% and 48% respectively). The coincidence factor is 62% for residential customers (compare with Table 3.9) whereas that factor is 92% for T32. That is connected to the fact that T32 has a peak at the same time as the aggregated peak.

Table 3.12 Results from USELOAD.

Tariff name *	Grid level (see figure 3.12)	Max load [MW]	Hour for max load (hour)	Utilization time (hours)	Energy [GWh]	Contribution to max load (%)	Conicidence factor (%)
T31	3	13.3	12	3767	50	2	82
H4	5	340.4	20	3399	1157	44	62
T32	5	249.3	10	3370	840	48	92
T33	3	20.1	8	4386	88	3	77
Aggregated	1		10	4609			

*The tariff names have the following meaning: T31 is a tariff for the service sector. T32 and T33 are tariffs for small and large industry. H4 is an ordinary household tariff.

3.4 Load management

Load research as it described in the previous sections, is dealing with passive observations of the load level and variation patterns. New technology however, opens up for possibilities, not only for passive observations, but for active load management.

By means of information and communication technology (ICT), it is possible to establish two-way communication with the consumers enabling frequent, automatic meter reading and direct intervention into the customers' use of electricity. This is included in the so-called smart grid technology.

In Norway, automatic meter reading (AMR) is going to be implemented by the end of 2016. That will entail hourly metering. Some distribution companies have already started the rollout.

Hourly metering makes it possible for consumers to follow and adapt to the running Spot Price. There is already a high and increasing share of Spot Price contracts in the Norwegian market, see Figure 3.17. As long as the metering and billing is based on conventional technology, Spot Prices have no impact on the load pattern. Instead of the actual hour by hour load, a standard load profile is used. It is therefore not interesting for a consumer reduce the load during high price hours. With hourly metering, the consumers will see and have the possibility to adapt to the Spot Price all the time.

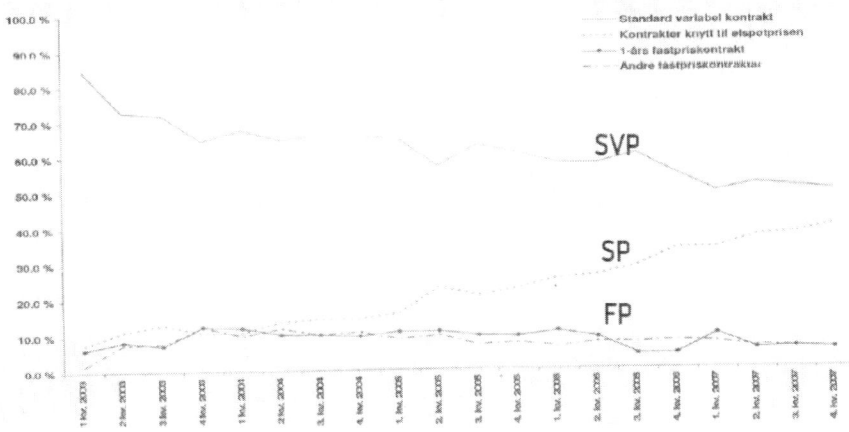

Figure 3.17. .Market share for contracts based on Standard Variable Price (SVP), Spot Price (SP) and Fixed price (FP) in Norway.

Direct communication with the customers makes control of individual appliances possible One example is load shifting of water heaters which have been tested in Norway. See Figure 3.18 .That can be done by the distribution grid company which is also responsible for the metering communication system. That type of control can be used to reduce peak load.as shown.

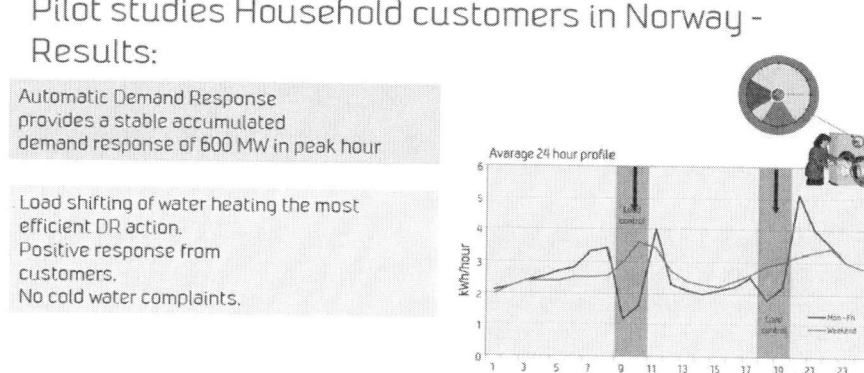

Figure 3.18.Demand response

3.5 References

[1] IEA: "Energy to 2050. Scenarios for a Sustainable Future" 2003

[2] WEC/IIASA: "Global Energy Perspectives" 1998

[3] Helle Grønli: "Prisfølsomhet overfor tidsvariable tariffer" EFI TR A4484, Januar 1997

[4] D. Hawden: "Energy Demand: Evidence and Expectations".Surrey University Press 1992.

[5] SINTEF Energy Research: " USELOAD Version 2.0, User Manual. TR F4821 February 1999.

[6] Klaus Livik, Nicolai Feilberg: "Energi- og effektforhold hos ulike kategorier sluttforbrukere", SINTEF Energiforskning TR A3998.

[7] Klaus Livik, Jan Foosnes, Nicolai Feilberg: "Estimation of annual coincident peak demand and load curves based on statistical analysis and typical load data", CIRED 1993.

[8] USELOAD: http://www.efflocom.com.

4 Pricing

4.1 Introduction

In the introductory chapter on microeconomics we concluded that the price should equal Short Run Marginal Cost (SRMC). That would maximize the economic surplus. In a perfectly competitive economy, that is achieved through price taking behaviour from the producers' and consumers' sides. Pricing, as it is discussed in this chapter, implies that somebody, i.e. a central decision-making institution, sets the prices. That was the case before deregulation. For the non-competitive part of the system, i.e. the grid, that is still the case.

In Norway and other countries where utilities were publicly owned, the authorities normally made the final decisions on electricity pricing before the restructuring. Municipal councils, county councils or the parliament had the final word. The pricing theory described here, had in many cases limited impact on practical decisions. This was partly because the political decision-makers had other priorities. Average cost pricing looked in many cases more attractive to voters than marginal cost pricing.

In countries where private ownership prevailed, for instance in the USA, the regulatory commissions had the mandate to decide the prices. In such cases, average cost pricing was also a normal outcome.

Under present circumstances, with a restructured electricity supply system, the electricity price is divided in two: One part covers generation, and is based on a market mechanism. The other part covers the transmission/distribution, and this price is based on central decisions. We return to this in Chapter 7 with more detail on transmission pricing.

The pricing theory described here is based on the assumption that the objective is to maximize social benefit, i.e. producer plus consumer surplus. But in practice electricity tariffs have to take several criteria into account in addition to maximum social benefit. These can be criteria such as:

- The tariffs should provide sufficient income to cover the costs for the utility.
- The tariffs should give a fair allocation of costs among the customers.
- The tariffs should be predictable and preferably stable in order to give the customers stable planning conditions.
- The tariff structure must be simple enough to make metering and billing possible in practice.

Such criteria can be inconsistent with the maximum social benefit criterion. In such cases a second best solution can be a good option. One example is Ramsey pricing described in Section 5.4.

In the electricity sector, as in certain other sectors, we are not only talking about a price, but a tariff. A tariff consists of a set of elements that determine the total charge. Traditionally the tariffs have been stable over certain period of time, typically one year. But the elements in the tariff can be variable over that time. Dynamic tariffs have been extensively used in some countries, for instance Germany and France. Dynamic pricing will be discussed in Section 5.3.

4.2 Marginal cost pricing

The principle of marginal cost pricing is that the price equals the cost of providing one additional unit. Then the consumers are faced with the true cost of the marginal unit, and this cost can then be matched against their marginal willingness to pay and thus lead to an efficient solution.

Depending on the time scale, the definition of marginal cost pricing will differ. We distinguish between Long Run Marginal Cost (LRMC) and Short Run Marginal Cost (SRMC). In this context long run means there is sufficient time to make investments. SRMC is the cost of increasing the production by a marginal unit without making any new investments.

But due to the special features of electricity, we can distinguish between different SRMCs based on the characterization the load (see Section 3.1). The "normal" SRMC is the cost of providing one extra kWh at a given instant of time, which is the relevant definition if we discuss spot pricing. But we can also define SRMCs for increased load over a certain period of time. One example is the SRMC for a standard or average household load profile. That is the cost of an extra kWh distributed over the year with that particular profile. Provided that the price (or the tariff) is kept constant over the year, this is the relevant SRMC for a household price.

Figure 4.1 shows how the instant SRMC develops as a function of output in a typical thermal power supply system. There is normally a considerable difference in SRMC between periods of low and high load, for instance between night and day in a thermal system.

Marginal cost pricing

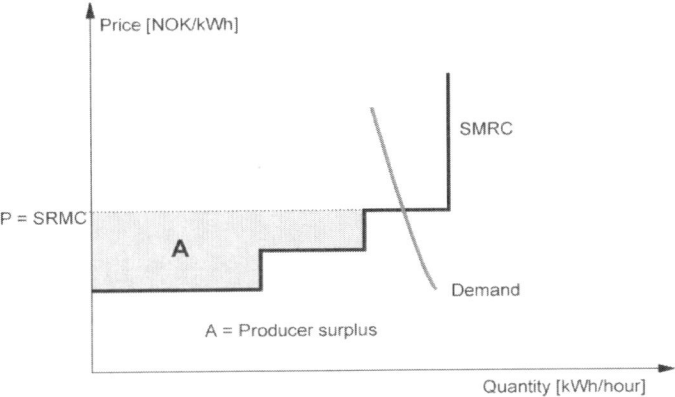

Figure 4.1 SRMC in a thermal generating system.

Figure 4.2 gives a simple illustration of how SRMC is affected by precipitation in a hydro dominated generating system. The precipitation has a strong influence on the SRMC. The difference between a wet and a dry year is substantial. Demand is also important, but not short term demand fluctuations. The cost of running hydro units up and down is small. There is no short-term cost for the water, but there is a limited amount available. The SRMC represents the value of having another unit of water available given the demand. Notice that quantity is measured in kWh/year so this is the SRMC with respect to a yearly load.

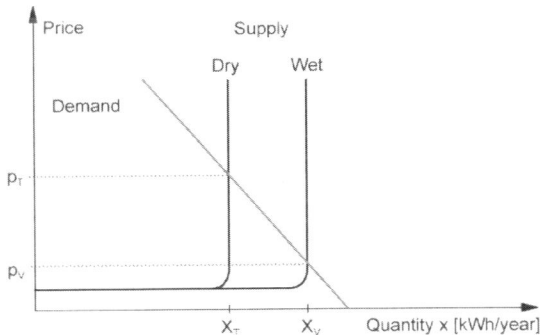

Figure 4.2 SRMC in a hydro dominated generating system.

The <u>instant</u> SRMC in a hydro dominated system is more difficult to estimate, but is based on the same principle. It is the value of one additional unit of water given the demand we anticipate. But it must be handled in a dynamic way because water can be stored in reservoirs. The instant SRMC (or the water value) is the expected value of a marginal unit of water given the present reservoir level, anticipated future inflow and future demand. System simulation models, for instance the EMPS model[1], make use of water value calculations. Through system simulations it is also possible to find SRMC for a yearly load. (Figure 4.2 is only meant to be a primitive illustration, not a basis for calculations.) Figure 4.3 shows the results of simulations with the EMPS model for the Norwegian power system as it was around 1990. What is calculated is the SRMC for an average load profile in Norway[2].

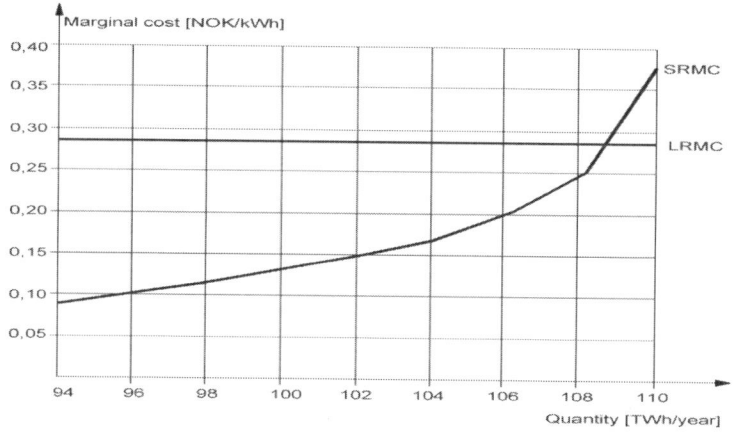

Figure 4.3 SRMC for the Norwegian power system.

The SRMC curve shown in Figure 4.3 was calculated by a number of system simulations with successively increasing consumption, all the time with the same load profile.

In Figure 4.3 we also show the Long Run Marginal Cost (LRMC) for Norwegian hydropower at the same time (1990). This is the cost of increasing production when the possibility of making new investments is taken into account. In his case it is shown as a constant, i.e. LRMC is not increasing as a function of quantity.

---

[1] The EMPS model is a system simulation model frequently used for the Nordic power system. The model is described in more detail in Chapter 6.
[2] The average load profile was earlier called the "firm power profile". It was the national load profile less power consuming industry and less occasional power deliveries. The term "firm power" has lost much of its meaning after restructuring. It is now more reasonable to call it an average load profile with power intensive industry excluded.

We will now discuss marginal cost pricing with the Norwegian situation around 1990 as an example. At that time, the power consumption in Norway was about 104 TWh/year. From Figure 4.3 we see that the corresponding SRMC was about 17 øre/kWh, which is considerably lower than the LRMC. Hence, the price of increasing the supply within the existing capacity limit was lower than doing so by investing in new generating capacity. This indicates an overcapacity in the system. With lower capacity we could have saved investment cost and the corresponding increase in operation costs would not fully outweigh that. But given this overcapacity, an optimal price is still SRMC , i.e. about 17 øre/kWh. In 1990 there was no common wholesale price, but the Statkraft Contract price represented a price level typical for the Norwegian wholesale market. At that time the Statkraft price, which was determined in the parliament (Storting), was about 22 øre/kWh (see Figure 4.3). A price drop from 22 to 17 øre/kWh, would increase the consumption slightly and thereby reduce the overcapacity to some extent. However, if we disregard this minor source of error, 17 øre/kWh would have been the right price at that moment. It leads to a correct balance between marginal willingness to pay and marginal cost, which in turn leads to a social optimum.

Out of this discussion it is possible to derive two general rules: one for the pricing and one for investments. The pricing rule is that the price should equal the SRMC. The investment rule is that in order to have optimal capacity, the expected SRMC should equal the LRMC. From this it is easy to see that with correct pricing and correct investments, prices in the long run should on average equal the LRMC.

It can be argued, however, that this is a short term and static way to look at the problem. The basic problem around 1990 was the overcapacity, and a useful way to correct the situation was to stop all investments and let the demand increase until the SRMC was equal to the LRMC. Obviously both the SRMC and the LRMC would represent optimal prices at that stage. However, it would probably have given unstable signals to consumers if the price was set to 17 øre/kWh and shortly after to its correct long-term level of 26 øre/kWh. With the consumption development in this particular time period, it would probably take only a couple of years for the demand to catch up with the overcapacity. A price setting rule that did not take these dynamic aspects into account would obviously be too simple.

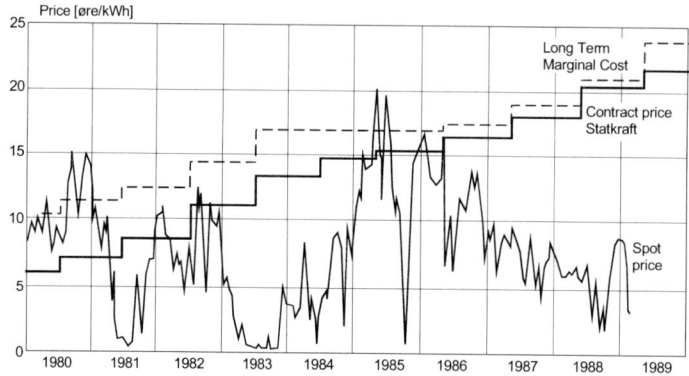

Figure 4.4 Wholesale prices in Norway before restructuring.

Figure 4.4 shows wholesale prices in the Norway during the last ten years before deregulation. There was a power exchange responsible for short term exchange between generating companies. The spot prices from this exchange reflected with a good approximation the SRMC in the generating system (see Section 5.9 for further details). The price was highly volatile. On average it was considerably lower than the LRMC, indicating an overcapacity in the system. There was no link between the spot price and the contract price.

4.3 Dynamic pricing

In general, dynamic pricing means price variability over time.

Dynamic pricing will always have some unwanted implications compared to a constant price. Metering and billing will be more complex and customers must follow price variations in order to adapt in a useful way.

Dynamic pricing does not necessarily mean that prices are unpredictable. Time Of Use (TOU) rates can follow a perfectly predictable variation pattern. Spot prices are more problematic, although they can be predicted to some extent.

4.3.1 Time Of Use (TOU) rates

TOU rates are rates that depend on time in a predetermined way. It can be time of the year (typical summer/winter), time of the week (typical week days/week end) or time of the day (typical day/night).

In order to explain the rationale of a TOU tariff, we start with a simple load duration curve and a generating system with variable SRMC. We assume variability of demand as shown in Figure 4.5. Low load is covering 50% of the time, high load is covering 40% and peak load 10% of the time. Figure 4.5 shows demand curves for these three levels of demand while Figure 4.6 shows the SRMC for the two types of generating capacity we have, base load and peak load capacity. In this case we assume that the demand during periods of low load is covered by base load units, while both peak and base load units are used to cover the demand in the remaining periods.

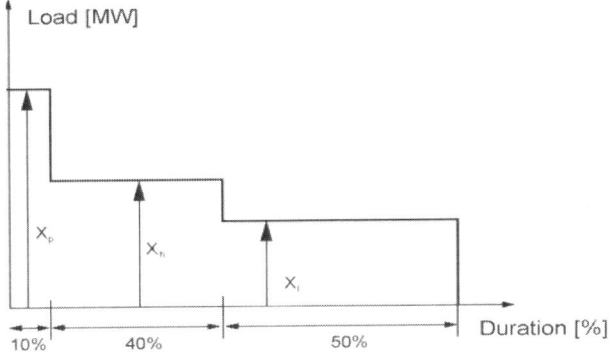

Figure 4.5 Load duration.

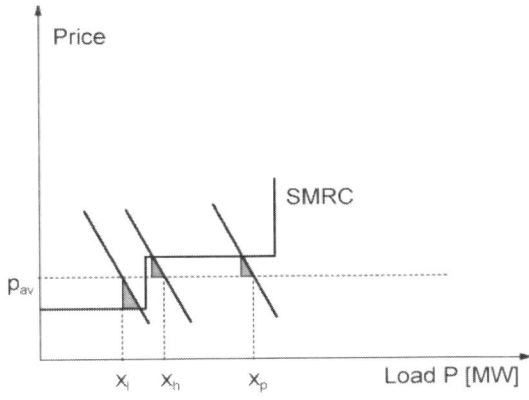

Figure 4.6 Impact of TOU rates.

We start with a constant price, p_{av} which reflects the average SRMC of the generating system. The constant price gives load variability as shown in the load duration curve. If we now introduce a TOU rate where the price follows the SRMC, the price will equal $SRMC_B$ under low load and $SRMC_P$ under high and peak load. This leads to an increase of the load, Δx_l, during periods of low load.

On the contrary it decreases by Δx_h and Δx_p, under high and peak load conditions. The corresponding gains in social surplus are indicated by the shaded triangles in Figure 4.6. The impact on the load duration curve is shown in Figure 4.7.

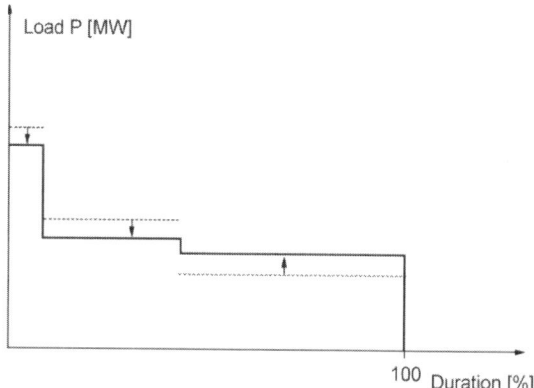

Figure 4.7 Load duration curve affected by TOU rates.

We have until now looked at TOU rates reflecting variable SRMC. However, TOU rates can also take investment costs into account, which is normally the case when we talk about *peak-load pricing*. [4]. The basic principle of peak-load pricing can be illustrated by an example which is even more simple than the one above.

Dynamic pricing

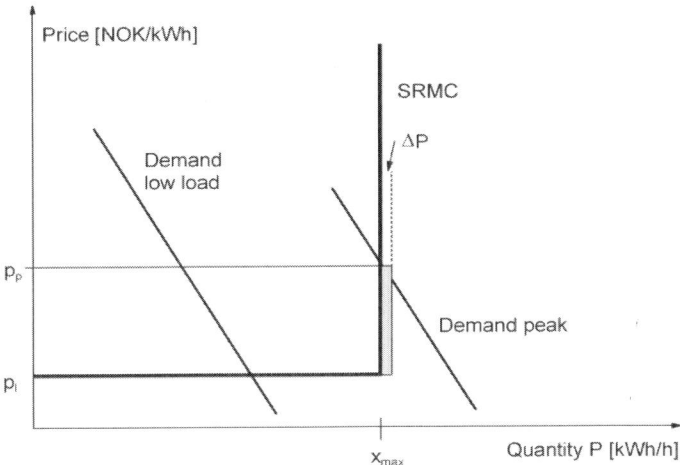

Figure 4.8 Peak-load pricing.

In this example we have a constant SRMC, implying that only one type of generating unit is available. We have only two different demand periods, i.e. low and peak demand. If sufficient generating capacity is available, the SRMC is therefore decisive for the price under both low and peak load conditions. However, during peak load the consumption may exceed the generating capacity. It is necessary with a raise in the price to keep the demand within this limit. The optimal peak price in the short-run is the price corresponding to the capacity limit. It is equal to p_p in Figure 4.8. If it is higher the capacity will not be fully utilized, if it is lower the capacity limit will be exceeded.

In the long run, the question is: should the capacity be extended? If a small amount of capacity ΔP is added (we assume the same type of units as before, i.e. the same SRMC), the social welfare surplus will increase by an amount equal to the shaded area in Figure 4.8 This additional surplus should be balanced against the investment or capital cost, which gives the following equation:

$$T_p(p_p - p_l)\Delta P = k_p \Delta P \qquad (4.1)$$

where:

- k_P is yearly capital cost for extending the capacity by one unit [kr/kW, year]
- T_P is the duration of the peak load [hours/year]

This gives for the peaking price:

$$p_p = \frac{k_p}{T_p} + p_l \qquad (4.2)$$

As an example we assume the following data:

- Yearly capital cost for new capacity: 500 NOK/kW/year.
- SRMC (fuel cost): 15 øre/kWh.
- Number of peak hours: 1000 hours.

The prices can now be calculated:

- Off peak load price: 15 øre/kWh
- Peak load price: 500 [kr/kW, year]/1000[hours/year] + 15 [øre/kWh] = 65 øre/kWh.

4.3.2 Spot pricing

Spot pricing means that the price should reflect the instant SRMC at the current spot in time and space. The spot price is variable on an hourly basis (or half hourly depending on the time resolution). It has to be calculated close to real time in order to reflect the current operational state and the marginal cost includes generation and network down to the spot where the customer is located.

Spot price calculation can be done by Optimal Power Flow (OPF). The procedure is described in Chapter 7.3. A spot price can come out of a market clearing process (which will be described in detail later), but here we will concentrate on a price based on central calculations and decisions.

As a by-product of central scheduling and dispatch, we will get the nodal prices $\lambda_i(t)$ at node i and time t reflecting the SRMC. In those nodal prices generation costs, network losses, network congestions and (provided Security Constrained Optimal Power Flow is used) security of supply constraints are taken into account.

"Spot Pricing of Electricity" [1] is a book which is frequently referred to, and is regarded as the theoretical basis of spot pricing. Terminology and notations in that book is somewhat different from what is used here: The spot price, $p_i(t)$, for a given customer i, at a given hour t, is defined as the sum of the following components [1]:

$$p_i(t) = \lambda(t) \quad \text{[System Lambda]}$$
$$+ \gamma_Q(t) \quad \text{[Generation quality of supply]}$$
$$+ \eta_{L,i}(t) \quad \text{[Marginal network losses]}$$
$$+ \eta_{Q,i}(t) \quad \text{[Network quality of supply]}$$
$$+ \eta_{R,i}(t) \quad \text{[Revenue Reconciliation]}$$

System Lambda represents marginal generation costs (provided there are no grid losses or restrictions between generators). Generation quality of supply and Network quality of supply represent restrictions that can be included in a Security Constrained Optimal Power Flow. Marginal network losses are also included. These components added together give a price reflecting SRMC. In addition a Revenue Reconciliation component is included. That means they are taking other objectives than economic efficiency into account. In order to include this Revenue Reconciliation with minimum loss of efficiency, Ramsey pricing is suggested.

4.4 Ramsey pricing

Ramsey pricing belongs to a class of rules or procedures that in economic literature is called *second best*. If, due to restrictions or other reasons, the optimal solution is unattainable, the problem is to find the best solution possible within these limitations.

Ramsey pricing can be applied when it is necessary to obtain higher revenue than a socially optimal pricing would give, e.g. in order to cover the utility's total cost. The objective is to depart from optimal pricing in such a way that welfare losses are minimized.

A condition for the use of Ramsey pricing is that market segmentation on the basis of price elasticity is possible. The Ramsey pricing rule, or the so-called inverse elasticity rule, says that the deviation in price relative to the optimum (which is usually marginal cost) shall be proportional to the inverse of the elasticity.

Ramsey pricing can be explained based on Figure 4.9.

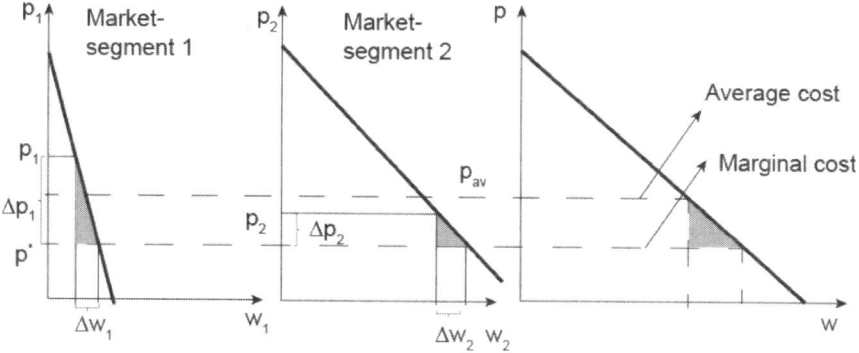

Figure 4.9 Different prices in different market segments

We have a market as indicated by the aggregate demand curve to the right, which can be divided in two market segments with different price elasticities. Segment 1 has a low elasticity, and segment 2 has a high elasticity. The production cost is indicated by the two dotted lines. The marginal cost, p^*, represents the optimal price, and p_{av} equals the average cost. If there is one uniform price in the whole market, it must equal the average cost p_{av} in order to cover the full cost. The corresponding drop in consumption compared to optimal price is Δx, and the welfare loss (deadweight loss) is indicated by the shaded area in the figure to the right. The question is then to what extent this welfare loss can be reduced by using different prices in the two market segments. The figure indicates a solution where the price in segment 1 is higher than the average cost while it is lower in segment 2. The corresponding welfare losses are indicated in the figure, and the sum of these two shaded areas is apparently smaller than the loss indicated to the right.

In order to find the total welfare loss, we start with the quantity effects in the two market segments.

Market segment 1:

$$\Delta x_1 = - \Delta p_1 \frac{\partial x_1}{\partial p_1} \qquad (4.3)$$

Market segment 2:

$$\Delta x_2 = - \Delta p_2 \frac{\partial x_2}{\partial p_2} \qquad (4.4)$$

The corresponding welfare loss can be found by calculating the area of the two triangles and adding them together.

$$Loss = \Delta p_1^2 \left(\left(-\frac{1}{2}\right) \frac{\partial x_1}{\partial p_1} \right) + \Delta p_2^2 \left(\left(-\frac{1}{2}\right) \frac{\partial x_2}{\partial p_2} \right) \qquad (4.5)$$

Our objective is then to minimize the loss expressed in Equation (4.5) subject to the condition,

$$x_1 \Delta p_1 + x_2 \Delta p_2 = K \qquad (4.6)$$

where K is a constant indicating a constant revenue. From Equation (4.6) we can find Δp_2.

$$\Delta p_2 = \frac{K - x_1 \Delta p_1}{x_2} \qquad (4.7)$$

We set this into Equation (4.5) and get:

$$Loss = \Delta p_1^2 \left(-\frac{\partial x_1}{2 \partial p_1} \right) + \left(\frac{K - x_1 \Delta p_1}{x_2} \right)^2 \left(-\frac{\partial x_2}{2 \partial p_2} \right) \qquad (4.8)$$

Minimum losses are found by setting the derivative with respect to Δp_1 equal to zero.

$$\Delta p_1 \left(\frac{\partial x_1}{\partial p_1} \right) = \left(\frac{K - x_1 \Delta p_1}{x_2} \right) \left(\frac{x_1}{x_2} \right) \left(\frac{\partial x_2}{\partial p_2} \right) \qquad (4.9)$$

Combining Equations (4.7) and (4.9) leads to:

$$\Delta p_1 \left(\frac{\partial x_1}{\partial p_1}\right)\left(\frac{1}{x_1}\right) = \Delta p_2 \left(\frac{\partial x_2}{\partial p_2}\right)\left(\frac{1}{x_2}\right) \quad (4.10)$$

Equation (4.10) shows in combination with (4.3) and (4.4) that in order to minimize welfare losses, the prices changes must give the same percentage change in quantity in the different segments.

$$\frac{\Delta x_1}{x_1} = \frac{\Delta x_2}{x_2} \quad (4.11)$$

The inverse elasticity rule can be derived by introducing the elasticity ε defined as:

$$\varepsilon(x) = -\left(\frac{\partial x}{\partial p}\right)\left(\frac{p}{x}\right) \quad (4.12)$$

By rewriting Equation (4.10) in the following way,

$$\left(\frac{\Delta p_1}{p_1}\right)\left(\frac{\partial x_1}{\partial p_1}\right)\left(\frac{p_1}{x_1}\right) = \left(\frac{\Delta p_2}{p_2}\right)\left(\frac{\partial x_2}{\partial p_2}\right)\left(\frac{p_2}{x_2}\right) \quad (4.13)$$

and introducing Equation (4.10), we get

$$\frac{\Delta p_1 / p_1}{\Delta p_2 / p_2} = \frac{\varepsilon_2}{\varepsilon_1} \quad (4.14))$$

which is the inverse elasticity rule.

4.5 Tariffs

A tariff has a set of elements. Figure 4.10 below gives an overview of the elements used in the Norwegian tariffs.

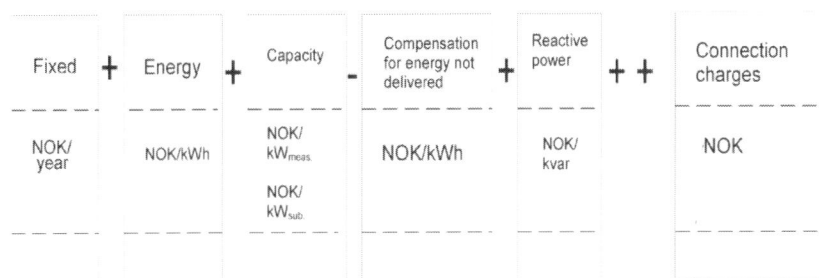

Figure 4.10 Tariff elements [2].

The fixed element is a neutral element which is independent of the customer's consumption and it has therefore no influence on the consumption pattern.

The energy charge is the consumption dependent element tied to the energy consumption. It can be constant or variable over time.

The demand charge (or maximum demand charge) depends on the demand at a certain instant of time (e.g. at the system peak-load) or at the peaking hour for the customer being charged. The demand charge can also be tied to a subscribed load level, which normally involves a surcharge or a penalty if the subscribed level is exceeded.

Compensation for energy not delivered. This is compensation to the customer for the inconvenience of supply interruptions.

Charge for reactive power is normally tied to the customer's maximum reactive load. It is normally not used for ordinary small customers.

Connection charges are one-time payments for being connected to the grid.

A tariff can be used for different purposes. It can make it possible to achieve targets which are difficult to reach with a linear price. It is for instance possible to use a fixed charge to achieve a revenue target, instead of using the more complex mechanism of a Ramsey-pricing.

In restructured power systems like the Nordic, linear prices are used for power purchases, while tariffs are used for grid services.

4.6 References

[1] F. C. Schweppe, M. C. Caramanis, R. D. Tabors and R. E. Bohn : "Spot Pricing of Electricity", Kluwer Academic Publishers, 1988

[2] H. Grønli and K. Sand: "Overføringstariffer – Prinsipper, struktur, mål og krav" EFI TR A4570, November 1997.

[3] Robert B. Wilson: "Nonlinear Pricing", Oxford University Press 1993.

[4] M. Boiteux: "Peak-load Pricing", Journal of Business, Vol. 33, no. 2 1960.

[5] M. A. Crew, C. S. Fernando, P. R. Kleindorfer: " The Theory of Peak-Load Pricing: A Survey" Journal of Regulatory Economics, pp 249 – 266, 1995

5 Restructuring/Deregulation of the Electricity Supply Industry

5.1 Terminology

The terms restructuring and deregulation are both used for the reorganization or transformation process of the power supply industry starting in many countries in the beginning of the 1990s. The term deregulation is most relevant in cases where the industry was privately owned and under public regulation which was typical for the USA. In those cases public regulation could be phased out for part of the system. In other cases, for instance where public ownership prevailed, the process was not really a deregulation. In such cases the extent of public regulation would in fact increase. Restructuring is therefore a more suitable term for this process, although it is a general term saying little about the content of the restructuring process.

The electricity supply industry is not the only one going through this type of restructuring. We see similar development in the telecommunications, gas supply and to some extent in the transportation sector. The restructuring of electricity supply is part of a worldwide liberalization process.

5.2 Economy of Scale and Unbundling in the Power System

Economy of scale in different parts of the electricity supply system is an important factor with respect to the restructuring. Economy of scale enables a company to produce with decreasing unit cost as the output increases. That makes it possible for one single company to supply consumers with a commodity at lower total cost than two or more competing companies. That is the basis for a natural monopoly.

It is generally accepted that electricity distribution is a natural monopoly. The cost per unit would be higher if two or more companies were to build and operate parallel distribution grids serving one specific area. The economy of scale for transmission is more questionable, but for several reasons (for instance operational coordination) it is most suitable to leave the responsibility for the transmission grid to one single company. Chapter 9 describes costs and regulation of grid companies in more detail.

Chapter 5: Restructuring/Deregulation of the Electricity Supply Industry

Electricity generation is assumed to have no significant economy of scale[1]. Several generating companies can operate in parallel without extra cost.

Due to these differences in economy of scale, restructuring of the power supply system had to include a division or unbundling of the system into two parts: one competitive part comprising generation and consumption, and one monopolistic part comprising transmission and distribution (T/D).

Open access to the grid is a vital presumption for efficient competition among the competitive parties. Generators and consumers must be able to use the grid for transportation and it represents a physical marketplace where trade can take place – on a competitive basis.

One step towards competition in the power supply industry was the opening of the grid to Independent Power Producers (IPPs). This was introduced in the US with PURPA[2] in 1978. Similar legislation was implemented in some European countries (Denmark, Germany, France, Spain and Italy), giving especially favourable conditions to production based on renewable sources or co-generation. These arrangements did not open up for competitive retail sales to consumers. The power supply company acted as a purchasing agency with monopoly on the transmission network and on sales to consumers. The term single buyer was in some cases used for this model.

A next step was the introduction of wholesale competition, giving distribution companies and large consumers the right to buy directly from producers. This gave open access to the transmission grid, but not to the distribution network.

The final step was to open up for retail competition.

Figure 5.1 illustrates the development. The shaded area – gradually decreasing as we move from left to right – indicates the increasing openness of the originally vertically integrated company.

---

[1] Economy of scale in the electricity generation sector is subject to some discussion. The generation cost per kWh is higher in a small unit than in a large one. But the difference has declined during the last years. The cost per unit is also decreasing as the size of a generating company increases due to management, financing, marketing, specialization and other factors. These factors apply to the electricity generation business in the same way as most other businesses. But, as in most other industries, economy of scale in electricity generation is not so extensive that competition is precluded.

[2] PURPA or Public Utility Regulatory Policy Act, was introduced by the Carter administration in 1978.

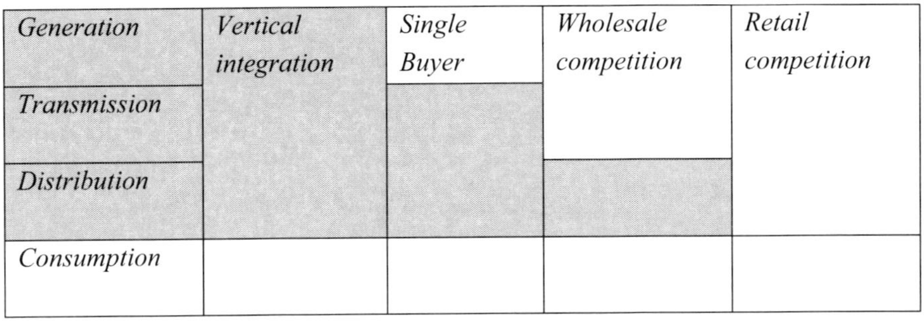

Generation	Vertical integration	Single Buyer	Wholesale competition	Retail competition
Transmission				
Distribution				
Consumption				

Figure 5.1 Different stages in grid openness.

5.3 Privatization

In some countries privatization of the electricity supply industry has been included in the restructuring process, but this is not always the case. In England and Wales privatization was an essential part of the restructuring. In Norway and the rest of Scandinavia, the public ownership was not affected.

To what extent competition can be combined with public ownership, is being discussed. Some will argue that an open competitive market must be based on private ownership. In the Nordic countries we have a long tradition state participation in industrial activity, not only in the power sector. Privatization was not an issue in the Nordic restructuring.

5.4 Stranded costs

Stranded costs in this context are costs caused by investments made before restructuring which cannot be recovered in the competitive environment after restructuring. These costs can be substantial, particularly in situations where heavy investments in for instance nuclear or hydro power plants have been done right before restructuring. If there is an overcapacity on the generating side leading to low market prices, generating companies can suffer substantial losses.

This is a serious problem in countries where the power supply industry was owned by private companies before the restructuring. Private industry will not tolerate economic losses if it can be avoided. In the USA, Investor Owned Utilities (IOUs) cover a large share of the market. This is also the case in Spain and some of the countries in Latin America. In these countries stranded costs were subject to a lot of discussion and different solutions were found.

In cases where the power industry was owned by public entities (municipalities or the State) before restructuring, no special provisions have been taken to recover stranded costs.

5.5 Roles and responsibilities in a restructured system

The roles and responsibilities of different companies and institutions in a restructured electricity supply industry are by no means a standardized. Different solutions are implemented in different parts of the world. In the present description we focus on the Norwegian and Nordic practice with some examples from other countries, especially England and the USA.

5.5.1 Regulator

The regulator represents the political authorities, i.e. the Government, and is responsible for ensuring that laws and regulations are followed and that public interests are taken into account. In the USA, the Regulatory Commissions have traditionally played an important role in controlling the electricity supply industry. There are regulating commissions on a federal (Federal Electricity Regulatory Commission, FERC) as well as on the state level. These commissions had complete control over electricity prices. They have also played an important role in the process towards deregulation in the USA. In the restructured industry the price control is limited to transmission and distribution services.

In England, a Director General of Electricity Supply (DGES) with extensive responsibilities was appointed as a part of the privatization process.

In its price control function, DGES is supported by the Office of Electricity Regulation (OFFER), which was established at the same time. OFFER has played an important role in the regulation of transmission/distribution prices. It later merged with a corresponding institution for gas into the Office of Gas and Electricity Markets (OFGEM), which is responsible for the regulation of gas and electricity. In promoting competition DGES is supported by the Monopolies and Mergers Commission (MMC).

The main regulating authority (Regulator) in the electricity sector in Norway is the Norwegian Water Resource and Energy Directorate (NVE). NVE is responsible for the monopolistic activities, which means the System Operator and the grid companies. NVE ensures that the Third Party Access (TPA) principle is fulfilled, e.g. the grid is available for any market actor without discrimination and under equal conditions. NVE also controls tariffs and revenues for the grid companies. The competitive activity within the industry, i.e. generation and supply, is controlled by the Norwegian Competition Authority. Basic objectives are to avoid collusive behaviour and to control and limit mergers and acquisitions, so that good market conditions are maintained.

Regulators in a restructured power system have the general responsibility for ensuring market efficiency. This includes control to avoid market power and ensure that there is proper treatment of security of supply. Supply interruptions can be regarded as an externality in the power supply sector.

5.5.2 Producers

Production companies are responsible for the generation of active power[3] and for selling it on the market. In many countries these companies used to have a monopolistic position, and thus had to be divided in the restructuring process to obtain efficient competition. One example is England were the former CEGB (Central Electricity Generating Board) was divided into three separate companies.

At an early stage, so-called Independent Power Producers (IPPs) were introduced in some countries. These companies came in as competitors to the established and in many cases monopolistic producers. In the restructured industry all generating companies can be regarded as IPPs.

Norway and the other Nordic countries had a large number of generation companies before the restructuring started. The largest company in Norway (Statkraft) had a market share of 30%. The largest in Sweden (Vattenfall) had 50% of the market. It was not found necessary to split up companies in this region. However, after the deregulation we have seen a number of mergers between generation companies, and in some cases the Competition Authority has interfered in order to avoid a too dominant market position for one single company.

5.5.3 Grid Companies

Grid companies are responsible for the operation and maintenance of the electricity network. They are also responsible for investments to connect customers to the grid and to maintain and, if necessary, extend the network capacity. One important responsibility for the grid companies in a restructured system is to keep the grid open for third party access and to work out and implement tariffs for use of the grid.

Metering and billing have always been part of a grid company's or a vertical utility's responsibility. Within the new system these activities have increased in complexity and it is no longer an obvious task for the grid company. It can for

---

[3] Generators can also generate reactive power, but that is not necessarily left only to the generators. The grid company can also produce reactive power. In Norway reactive power is not subject to market competition. The grid companies have the exclusive right to buy and sell reactive power.

instance be left to the supplier. However, according to Norwegian regulations, metering and dissemination of metering data is the responsibility of the grid companies. Metering data is necessary for the supplier as well as the grid company and the data should be consistent.

Transfer of electric power can be combined with transfer of data. Signals can even be transferred on the same wire. There is normally some economics of scope that make communication an attractive activity for a grid company [4].

5.5.4 End users

The power consuming part of the system has traditionally been described as "non-dispatchable". Consumers can be affected by pricing, or more generally by tariffs, but not by direct control. As indicated in Figure 5.1, the consumers are not part of any vertically integrated power system. They are independent in the sense that they decide their own consumption pattern (within certain limits).

However, modern information and communication technology offers improved possibilities of direct control of individual end users. Combined with advanced tariffs, this can be used to monitor and control consumption and thereby improve the efficiency of the system.

A basically new feature introduced with the restructuring, is the consumers' free choice of suppliers. In some cases all consumers are eligible, i.e. they have this freedom. In some cases eligibility is not fully implemented or it is implemented step by step.

With respect to end users' flexibility in the market, metering equipment is important. Meters with hourly recording makes it possible to buy electricity directly in the spot market (note that this is also possible in a rough manner by use of standard profiles, as will be explained later), and it is easier to change supplier. Hourly recording, in many cases combined with two-way-communication, is on its way into the system.

---

[4] In economic terms we talk about *economics of scale* and *economics of scope*. Both concepts are relevant for a grid company. As pointed out in Section 5.2 economics of scale means that unit costs are decreasing as the size or the scale of a company increases. In our case it means that T/D cost per kWh decreases as the number of kWh/year increases. Economics of scope means that combined production of two commodities in one company will cost less than producing the same commodities in two different companies. In our case it means that to build and operate a data communication and a power network will cost more if it is done by two different companies than if it is done by one.

5.5.5 Power Exchange (PX)

The Power Exchange or Market Operator is responsible for receiving bids for sales and purchases of electricity, and to match the bids in such a way that prices and quantities are settled. The basic activity of the power exchange is operation of the short term physical electricity market, the spot market. The spot market is normally a day-ahead market with a time resolution of 30 minutes or one hour.

The Nordic Power Exchange, Nord Pool, is the only common market place for the Nordic power system. Nord Pool ASA is registered as a Norwegian company and has to adapt to the regulation of the Norwegian authorities. Nord Pool is responsible for both physical and financial trade.

Power Exchange or Power Pool

What is the difference between a pool and an exchange? The two concepts are closely related, but there are certain features that are typical for each of them.

Power pooling institutions existed long before deregulation came along. We had a Norwegian pool organized by the former Samkjøringen and we had different pools in the US. These pooling institutions were established in order to make better use of existing production resources. A central scheduling and dispatch (operation optimization) was normally performed by the pooling institution and a dispatch centre could be included. This optimization resulted normally in some power exchange, but that was exchange between generating companies. The objective was pooling i.e. optimization of resources, not trade.

The focus of a power exchange is trade. It is a market place. We know from economic theory that a perfectly free market also brings about optimal use of resources. A power exchange will normally not include an explicit optimization. But if the conditions for perfectly competitive market are met, the outcome will be the same.

According to this definition, Nord Pool is a power exchange, and not a pooling institution. But the borderline between the two terms is by no means clear.

5.5.6 System Operator (SO)

The basic responsibility of the System Operator is security of supply. As mentioned above, supply interruptions represent a serious externality in the power system and the SO is responsible for taking necessary action so that an efficient solution is obtained. The SO is responsible for hour by hour – and to some extent minute by minute – power balance and for keeping sufficient capacity margins in the generating system and the grid. This is done in order to keep security of supply at an acceptable level. In addition, the SO is responsible for keeping frequency variations within acceptable limits and for control of the voltage.

The System Operator can either be an independent entity – a so-called Independent System Operator (ISO), which is normal in the US – or system operation can be integrated with the responsibility for the transmission grid. These entities with integrated responsibilities are called Transmission System Operators (TSOs), and are found in most European countries. In the Nordic area there are four TSOs which own and operate the main grids in addition to being System Operators. Extensive coordination and cooperation between the Nordic TSOs has developed.

Some of the tasks of the Power Exchange and the System Operation are closely related and both entities can be involved. Congestion management is one example. In the Nordic region, transfer limits are set by the SOs. Nord Pool plays a major role in relieving congestion between price areas (see section 7.2), while the TSOs are responsible for congestions within price areas.

In some cases the PX and the SO are integrated into one unit. This is especially relevant in cases where operation is based centralized scheduling/dispatch. The English solution before 2000 was based on central scheduling/dispatch. The PJM solution is also based on a common PX/SO institution.

5.5.7 Balance responsible entities

This type of entity is responsible for working out a balanced hour by hour schedule, normally for the day ahead, and for keeping the real-time balance close to that schedule. Normally there is a penalty for deviations from that scheduled balance.

The tasks of the balance responsible entities are depending on the market design. In California the so-called Scheduling Coordinators (SCs) used to play an active role in the system. The California Power Exchange (CalPX) was one of the SCs[5].

In the Norwegian market all generators and consumers have to be tied to a balance responsible entity. Large consumers, typically power consuming industry and generating companies are balance responsible themselves. Small consumers and producers are normally using other entities that are responsible for their balance.

The balance responsible entity presents a balanced schedule for the SO. The schedule can be based on a mix of generation, consumption, spot sales or purchases and bilateral contracts. Any deviation from the balance presented in the

---

[5] The California electricity market went through a major crisis in 2001 and the Power Exchange as well as other entities went bankrupt. In the aftermath of the crisis there has been a major redesign of the market in California

schedule, will be compensated through the Balancing Market. More details about that in Chapter 9.

In order reduce the dependency on the Balancing Market, the balance responsible entities can use the Elbas Market, which is organized by Nord Pool, to adjust the balance until two hours before real time. Elbas is based on continuous auctioning.

5.5.8 Suppliers

Suppliers are market participants that sell electricity to end users. Suppliers for small consumers (typically without hourly metering) are often called retailers. The majority of the suppliers are generating companies or traders. A grid owner can also run a supply business, but according to Norwegian regulations it has to keep separate accounts for the grid and the supply businesses. A supplier will normally be a balance responsible entity. If not, that responsibility has to be carried by another entity.

The local distribution (or distribution/generation) company, the former monopolistic utility, is normally the dominant local supplier. In Norway 80-90% of the customers are still buying electricity from their local supplier.

5.5.9 Traders and Brokers

A trader is buying and selling on the electricity market. A trading company can be engaged in physical and/or financial trade. Physical trade requires balance responsibility, financial trade does not.

A broker is not engaged in trading, but it is establishing contact between buyers and sellers. Brokers are in many cases defining standard contracts used by market participants.

5.5.10 Market Participants

This is a general term used for entities that are active in the electricity market, normally on the wholesale level. Other terms like marker player and market actor are also used. These can be producers, consumers, traders or brokers (operating on behalf of others) taking active part in the market process.

In many cases one company can be engaged in many of the roles described above. A large company like Statkraft is of course a power generation company. At the same time it is also a market participant, a trader (Statkraft is very active in the financial market), a supplier, and a balance responsible entity.

5.6 Market design

A free and open market requires that a large number of buyers can buy from a large number of suppliers or producers. In an efficient market all participants must have equal access to the market and to all relevant information about prices and supply conditions. It is necessary to have information about both the present state and the factors affecting future conditions.[6]

Market design means the rules and practical arrangements governing how the different entities operate. The main objective is normally to obtain an efficient market, but, as we have seen in some cases, this is not always achieved.

5.6.1 Market access

Full access to the market for all entities (wholesale and retail) from both the supply and demand side is a necessary condition for an efficient market. Full market access on the retail level requires not only legal access, but also suitable practical arrangements concerning metering and billing. Practical arrangements must be simple and inexpensive, and the charge for a change of supplier must be low or zero.

In most countries full market access was not implemented in one step. In England the first step was taken in 1990 when the market was opened for consumers with consumption above 1 MW. In 1994 the limit was lowered to 100 kW, and finally in 1998 full access was implemented.

In Germany full legal access was introduced from the start, but difficulties concerning metering and billing as well as other practical obstacles have so far limited consumer mobility.

---

[6] One important factor for present and future electricity prices in a hydro power dominated market as the Nordic one, is the water content in the reservoirs. At an early stage in the in the liberalization process, information about reservoir levels was kept confidential by each generating company. This created an asymmetric information situation in the sense that large generating companies with several hydro power plants had more information than small ones. In addition, generating companies in general had more information than the consumer side. This asymmetry of information was deemed harmful by the regulating authorities and generating companies were instructed to make the information available. Statistics Norway was given a mandate to collect and disseminate information about hydro power reservoir content on a regular basis. Another example is scheduling or outages of different components in the system for maintenance and repair. If, for instance, a nuclear plant is taken out for maintenance, this will have impact on the market price. This type of information is also made generally available in the Nordic market. It is reported to Nord Pool, which makes it available on the Internet.

Full market access for producers is generally included in restructured systems. For small or distributed generators market access is in some cases supported for environmental or other reasons.

5.6.2 Exchange/pooling

Different pools

An exchange or pooling institution is a crucial element in a restructured electricity system. There are different ways in which such institutions operate. Various procedures for price settlements exist, and the relationship between the exchange institution and its users can also be organized in different ways. A comparison between Nord Pool and the (pre-2001) English Pool can serve as an illustration. See Figure 5.2.

The English Pool had a monopolistic position in the sense that all physical trade should go through the Pool. The Nordic Pool has always been voluntary. Presently about 70 % the physical trade in the Nordic area goes through the Nord Pool Spot. The volume traded on the Nord Pool Spot has been increasing year by year.

The spot exchange or the pool will normally be in a monopolistic position in the sense that there will only be one entity engaged in physical spot trade in a given area. In Norway the last revision of the Energy Act introduced a legal monopoly for spot trade. In other countries the monopolistic position is not necessarily enacted by law, but in most cases there is a de facto monopoly. One exception is India where we find two competing exchanges, one in New Delhi and one in Mumbai.

There are strong reasons to have only one physical spot exchange. One argument is that competing market places will each have a weaker liquidity compared to the monopoly, which in turn can increase uncertainty, volatility and potential market power. For financial trade, which is based a spot price as a reference, it is preferable to have only one spot price to refer to.

Outside the physical spot market, there can be several competing market places.

Price settlement

There are different procedures for the settlement of prices and quantities in a pool or a power exchange. Generally we can distinguish between two broad categories of price clearing processes. The first is *periodic clearing* and the other is *continuous trade* or *continuous auction*.

Chapter 5: Restructuring/Deregulation of the Electricity Supply Industry

Periodic clearing.
Definition:
The whole clearing process is done in one operation and repeated typically once a day
Characteristics:
One-time bid or two (perhaps more) repeated bids. All receive clearing price (SRMC)
Multilateral trade
Examples:
Full scheduling and dispatch (Former English pool, PJM)
Price cross (Nord Pool Spot, EEX Germany, APX the Netherlands)

Periodic clearing is a bidding and price clearing process where all the information is collected through one or a limited set of bids and all computation and distribution of information to the involved parties is done in one single operation. Nord Pool uses one round of bids, but it has been discussed to introduce repeated bids, especially in cases where sales and purchase curves do not intersect.

The process is repeated with regular intervals, normally once a day.

Another characteristic of the periodic clearing is that all participants receive the same price, which is the short term marginal cost (SRMC)[7].

Periodic clearing, which is most common in the physical spot market, can take two alternative forms based on either centralized or decentralized scheduling.[8] Centralized scheduling resembles much of the old type of coordinated scheduling in the monopolistic power system (see Chapter 6 for more details). All costs and restrictions of each generating unit are taken into account and a complete schedule (unit commitment and dispatch) for the entire generation system is worked out. This is an extensive and complex procedure. The former English pool is one example. The basic difference compared with the old system is that the data needed for optimization is provided through bids. Centralized and decentralized scheduling exemplified by the English and Nordic solutions is illustrated in Figure 5.2.

---

[7] The short term marginal cost (SRMC) which, in an optimally dispatched system, is the same for all units (see Chapter 6 for a more detailed discussion) is called the system marginal cost (SMC) or sometimes *system lambda*.

[8] Different terms are used. In [5] the terms *integrated and unbundled* are used.

Market design

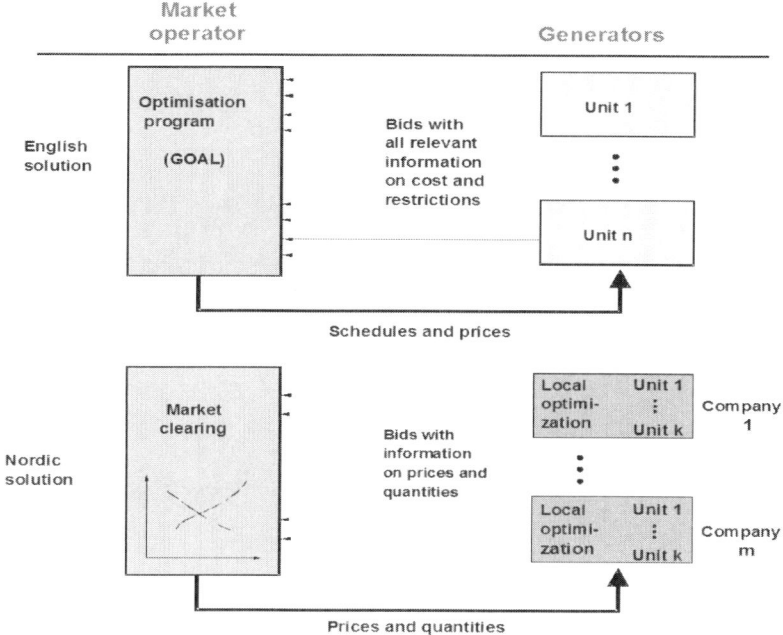

Figure 5.2 Centralized (England and Wales before 2001) and decentralized scheduling (Nord Pool).

Production schedules and prices (SMCs) are the outputs from the optimization program. The fact that these prices are decisive for the income of the companies (in contrast to the old system where these prices had minor impact on the economy of the company) makes the bidding vulnerable to market power in the same way as bidding in a simpler price clearing system.[9]

In a decentralized system, the market clearing price and traded quantity are calculated simply by finding the intersection point for the sales and purchase curves. The price calculation is normally not taking intertemporal links into account. The bids are aggregated and the clearing price is found for one hour at a time without regard to the hour before or the hour after. It is left to the generating companies to include intertemporal links in the scheduling. This scheduling has to be done before the biding starts and the prices are unknown at that stage. The scheduling therefore has to be based on price forecasts. This decentralized

---

[9] Experience from the English pool indicates clearly that market power has been a problem. That has probably most to do with the fact that there were only very few generating companies in the system.

scheduling deals with factors such as start-up and shut-down costs, minimum uptime, minimum downtime and maximum ramping rate[10].

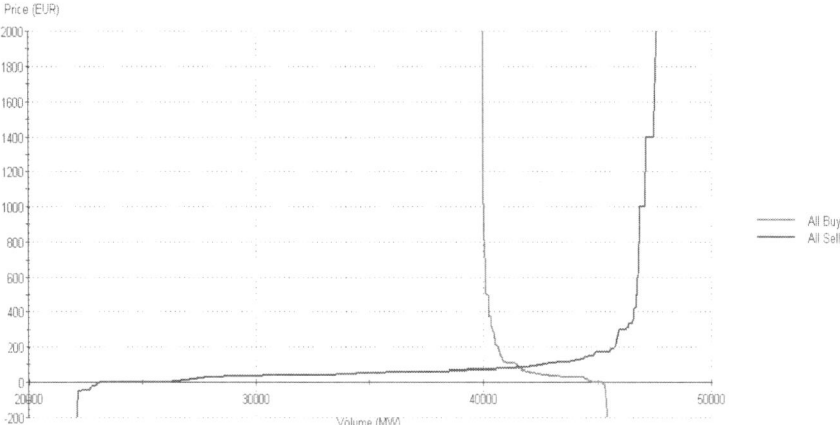

Figure 5.3. Example from Nord Pool. Buy and sell curves for hour 4, February 22, 2010. Price: 78.85 Euro/MWh

In a system based on central scheduling, the different intertemporal ties can be handled directly in the optimization. All costs and restrictions can be included, also costs and restrictions relating to the grid. In this case the optimization will be complex.

In order to improve the planning framework for generating companies in a decentralized system, Nord Pool has introduced *block bidding*, which means that the companies can tie the bids for a predefined number of hours together into a block. Either the whole block is accepted or it is rejected in the market clearing.

---

[10] Generation planning in general is described in Chapter 6.

Market design

Continuous auction
Definition:
Continuous clearing process
Characteristics:
Pay-as-bid price
Bilateral trade
Decentralized scheduling
Examples:
NETA,
Derivatives market in NASDAQ OMS, Elbas in Nord Pool Spot

A continuous auction is a process where open bids are displayed on the "marketplace" and potential buyers and sellers can pick offers that look attractive. In such an auction a bidder must accept the price it has offered. It is a *pay-as-bid* price, in contrast to a periodic clearing where the price generally will differ from the price offered by the bidders. In a price clearing system only the marginal bidders will receive a price equal to the bid-price. Every bidder can therefore reveal its marginal cost or marginal willingness to pay without risking an unfavourable price. This will not be the case in a pay-as-bid auction. Every bidder will try to obtain a price close to (or even better than) the market-clearing price.

Continuous trade is normally based on bilateral transactions.

A continuous auction requires decentralized scheduling. It is not possible to run any form of central scheduling as long as long as physical trading transactions are done one by one. Scheduling and trading will be an integrated in a more or less continuous process.

The auction will close a number of hours before real time. A continuous auction will normally go closer to real time (two hours typically) than a periodic clearing process. After the market is closed it is left to the PX and/or the SO check for feasibility. If constraints are violated, adjustments have to be made. This can be done through the balancing market.

5.6.3 Reserves and balancing

The spot market is an ex ante market. It closes some hours before real time. In order to make sure there is physical balance in real time, it is necessary to have reserves available. It is the responsibility of the SO to make sure there are necessary reserves available and to deploy the reserves if and when needed.

The need for reserves is to some extent affected by the market design. If it is possible to adjust the market balance closer to real time (which is possible through the Elbas market in Nord Pool) the requirement for reserves can to some extent be reduced. But it is always necessary to have reserves available in case of contingences.

We distinguish between different types of reserves. Response time, activation mechanism (automatic or manual) and other characteristics are different. Reserves are included in a broader set of supporting services called ancillary services. It is described in more detail in Chapter 8.

There are different solutions to the problem of making such reserves available. We can broadly distinguish between centralized and a decentralized solutions. In the first case the SO takes the responsibility for acquiring what is needed and (if necessary) for remunerating the companies for making the reserves available. In the second case the SO defines the requirements and distributes the obligations for reserves among the market participants. It is then left to the participants to rely on self-provision or to buy reserves from others in order to fulfil their obligations.

There is normally a cost for keeping the reserves ready for deployment and another one for their use. These costs are different for different generating units. Generally hydro power units have lower costs than thermal units. A description of how reserves and balancing is handled in the Nordic system will be given in Section 5.8 and more extensive discussion about ancillary services in general will be given in Chapter 8.

5.6.4 Physical and financial trade

The spot market and the balancing market are both physical markets[11]. Normal retail sale from a supplier to a customer is also physical. The contracted quantity is delivered at the agreed upon location (area or node), time and price. A financial contract will also include a specification of quantity, location, time and price, but there is no physical delivery included, it is only a financial transaction.

---

[11] Definitions and terminology yare different in different parts of the world. In the Nordic system, the spot market is regarded as a physical market despite the fact that it is day-ahead market. The quantities traded are based on the bidding and price clearing system not on the quantities metered in real time. The balancing market on the other hand is a true physical market. In the PJM system, The Day-Ahead Market is (correctly) regarded as a forward market in which hourly LMPs are calculated for the next operating day based on generation offers, demand bids and scheduled bilateral transactions. In addition there is a Real-Time Market in which current LMPs are calculated at five-minute intervals based on actual grid operating conditions. These Real-time prices are made available and PJM settles transactions based on these prices.

Financial trade requires a reference price against which the financial transactions can be settled. The most common reference price is the spot price.

Example: A consumer enters into a financial contract for one kWh at a given location and time with the spot price as the reference. If we assume that the contract price is 20 øre/kWh and the spot price at delivery time is 30 øre/kWh, the consumer will be paid 10 øre/kWh from the seller. If the spot price is 15 the customer has to pay 5 øre/kWh back to the seller. In both cases the customer will be confronted with a net purchase price of 20 øre/kWh if the customer decides to buy his kWh on the spot market.

One of the advantages of using a financial contract (instead of a physical one) is that the consumer always will see and adapt to the running spot price. If the spot price is extremely high, the customer might chose to lower consumption and put the savings in the bank. A spot price of 50 øre/kWh means that he/she will earn 50 øre for every kWh saved. A physical contract will normally be based on measured consumption which means the customer would save only 20 for every kWh of reduced consumption.

5.7 Restructuring in different countries

An early initiative towards privatization and market introduction was taken in Chile in the late 1970s. Other parts of Latin America (Argentina, Peru, Brazil Colombia) have to some extent followed during the 1990s.

In Europe the pioneer was England and Wales, where privatization and deregulation came with the Electricity Act of 1989. The state owned company Central Electricity Generating Board (CEGB) was broken up in three generating companies: National Power, PowerGen and Nuclear Electric and one transmission company was established: the National Grid Company (NGC). 12 regional distribution companies (RECs) where established. The RECs served as distribution grid companies as well as suppliers. All these companies were privatized expect Nuclear Electric which maintained state ownership until 1996.

A power pool was established, operated by the NGC. It was a mandatory pool based on bids from the generators (not consumers) and a central scheduling/dispatch algorithm as described earlier. This was replaced by the so-called New Electricity Trading Arrangements (NETA) in 2001which was based on continuous trading. A third reform led to the establishment of the British Electricity Trading and Transmission Arrangement (BETTA), in operation from 2005. That includes Scotland.

The fact that only three generating companies were to compete in the market, made *market power* an obvious problem in the UK. Nuclear plants are largely excluded from competing due to its low marginal cost, so we are actually left with a duopoly. After some time, however, new entrants came in, investing in CCGT[12] plants. That increased the number of competing companies.

The market opening on the consumer side was introduced stepwise in the UK. At the outset only consumers with peak load above 1 MW were eligible, i.e. free to choose supplier. That limitation war reduced to 100 kW in 1994 and completely removed in 1998, making all consumers eligible.

In the Nordic region, Norway was the pioneer. The Energy Act of 1990 opened up the Norwegian market. In contrast to the UK, all consumers were eligible right from the start, but there was a fee for switching supplier. That fee was war removed in 1997. The other Scandinavian countries and Finland joined this market during the 1990s. (More about the Nordic development in Section 5.8).

Liberalisation in the rest of Europe has largely been a top-down process driven by directives from the EU. The directives lay down the principles and conditions for creation of a single Internal Electricity Market (IEM). The first one issued was the Internal Electricity Market Directive in 1996 (96/92/EC) that set goals for a gradual opening of the electricity market for all member states. A new Directive in 2003 (54/EC) brought in some adjustments. The goal was to open the market for electricity (and gas) by 1 July 2007.

Unbundling has been implemented and TSOs are established, normally one in each country (except Germany and the UK with four each) with responsibility for the transmission grid and for system operation. At the moment, a set of power exchanges are in operation in Europe: In the Nordic area (Nord Pool), in Germany (EEX), in France (Powernext) and in addition Italy, Poland and Spain have national power exchanges. All these exchanges are based on the simple price cross algorithm and are independent of the TSOs.

The European Commission founded the Council for European Regulators' Group for Electricity and Gas (ERGEG) which, together the voluntary Council of European Regulators (CEER), are promoting an integrated IEM. In addition there is a co-operation between the system operators through the association of European Transmission System Operators (ETSO), presently ENTSO-E. The former NORDEL, a similar organisation covering the Nordic countries, entered into ENTSO-E in 2009.

These institutions are working for a closer integration of the European electricity market through for instance *market coupling*. See 7.4 for more details. Another topic subject to much attention during the last few years is *inter TSO compensation (ITC)*. That means compensation to one TSO for transfer of power through its grid between two other TSOs. Different principles are being discussed.

In New Zealand the liberalization started in 1994. The wholesale electricity market (NZEM) was reformed and competition was introduced through the creation of a

---

[12] Combined Cycle Gas Turbines

spot market for electricity. The transmission system is owned and operated by a state-owned enterprise, Transpower, which performs the function of grid-owner, system operator, scheduler and dispatcher for the wholesale market. Prices and quantities are determined half-hourly at each node in the transmission grid. That is the basis for spot trading in NZEM, which starting in 1996. NZEM is considered to be the first electricity market based on nodal pricing. It is therefore a predecessor for the locational marginal pricing models (LMP) widely adopted in the USA later. [10]

In the USA, restructuring legislation began with the Energy Policy Act of 1992 which launched a national effort to restructure the electricity supply system and open for reliance on markets. The Federal Energy Regulatory Commission (FERC) took the lead in in Order 888 by opening access to the transmission grid. That triggered the formation of Independent System Operators (ISOs) including California ISO (CalIso) New York ISO (NYISO) etc. Later the FERC Order 2000 encouraged the formation of Regional Transmission Organisations (RTOs). [10], [11].

The California crisis in 2000/2001 slowed down the process in the USA, but not for a long time. In California a power exchange (PX) had been established separate from the ISO, and a zonal pricing system (area prices) introduced, more in line with the Nordic system than with other parts of the US. Retail access was approved, but most retail prices were frozen. This had to do with the recovery of stranded costs. All revenues that exceeded costs, but fell below the frozen price level, were used for stranded cost recovery. This retail price freeze excluded demand response and made the system vulnerable to market power. In addition the suppliers were exposed to high risk. The suppliers had to buy at an unregulated wholesale price and sell at a fixed retail price. As long as the wholesale prices were low, no problems appeared. But in the summer 2000 the prices started to grow and eventually skyrocketed, leading to what is later known as a "melt down". The California "melt down" included bankruptcies and rolling blackouts. Eventually the state government had to take over major parts of the electricity business in California. [11].

In other parts in the US, markets without similar design flaws were introduced. In many cases LMCs were used. The integrated PXs/ISOs took the responsibility of system and market operation including calculation of LMCs based on Security Constrained Economic Dispatch (SCDP)[13]. A Standard market design (SDM) was developed. The system in Pennsylvania-New Jersey-Maryland (PJM) is one example of a well-functioning electricity market. The market uses locational marginal pricing (LMP). If the lowest-priced electricity can reach all locations, prices are the same across the entire grid. A DC-equivalent model is used, see section 7.3.9. When there is transmission congestion, energy cannot flow freely to

---

[13] A simple description SCDP is found in section 7.3 7

certain locations. In that case, more-expensive electricity is ordered to meet that demand. As a result, the locational marginal price is higher in those locations.

The Energy Market consists of Day-Ahead and Real-Time markets. The Day-Ahead Market is a forward market in which hourly LMPs are calculated for the next operating day based on generation offers, demand bids and scheduled bilateral transactions.

PJM Interconnection auctions Financial Transmission Rights (FTRs) to assist market participants in hedging their price risk when delivering energy on the grid.

FTRs are financial instruments that entitle the holder to a stream of revenues (or charges) based on the hourly congestion price differences across a transmission path in the Day-Ahead Energy Market.

FTRs provide a hedging mechanism that can be traded separately from transmission service. Market participants are able to hedge against their congestion costs by acquiring FTRs that are consistent with their energy deliveries.

PJM Interconnection currently operates two markets for *ancillary services*:

Synchronized Reserve supplies electricity if the grid has an unexpected need for more power on short notice.

Regulation is a service that corrects for short-term changes in electricity use that might affect the stability of the power system.

In addition, Black Start Service supplies electricity for system restoration in the unlikely event that the entire grid would lose power.

More about ancillary services in Chapter 8.

Demand response (also known as load response) is end-use customers reducing their use of electricity in response to power grid needs, economic signals from a competitive wholesale market or special retail rates.

Demand response is an integral part of PJM's markets for energy, day-ahead scheduling reserve, capacity, synchronized reserve and regulation. Demand response can compete equally with generation in these markets.

In PJM's Energy Market, end-use customers participate in demand response by reducing their electricity use either during an emergency event or when locational marginal prices (LMPs) are high on the PJM system. End-use customers participate in demand response in PJM through members called curtailment service providers (CSPs), who act as agents for the customers. A list of PJM members who are CSPs is available [12]

In Asia, Eastern Europe and the former Soviet Union, restructuring is being discussed and in some countries important steps are taken.

India introduced power system restructuring with the energy act of 2003. The Central Electricity Regulatory Commission (CERC) introduced open access to the transmission grid from May 2004, facilitating trading from one utility to another. But open access is not introduced on the retail level. Responsible for the operation of the system are the National Load Dispatch Centre (NLDC) as a co-ordinating institution and four Regional Load Dispatch Centres (RLDCs). Two competing power exchanges are established, one in New Delhi and one in Mumbai. Turnover is so far low. Trading is to a large extent left to bilateral transactions. One interesting detail is that prices for so-called unscheduled interchange (similar to balancing power in the Nordic system) are linked to system frequency. Low frequency, high price and vice versa. [13]

5.8 The Norwegian and Nordic restructuring

5.8.1 Introduction

The restructuring was introduced in Norway with the Energy Act of June 1990, which was effective from January 1991. It took some time to establish the necessary organizational structure. These arrangements were completed in May 1992 and since then there has been an open electricity market in Norway.

In Sweden the process towards an open trading system took several years. The final decision was taken in October 1995 and the open Swedish/Norwegian market, the first electricity market completely open to trade across national borders, has been in operation since January 1996.

In Finland a similar restructuring has been done. Finland joined the open Swedish/Norwegian market in 1998.

In Denmark it took some more time. Denmark is divided in two separate grid areas, where Jutland/Fyn is connected to the Continental grid and Sealand to the Nordic grid. Both areas joined the open Nordic market in 2000.

The Norwegian reform did not include any change of ownership. In contrast to the English situation, no privatization was proposed or even discussed. The Norwegian power industry has been dominated by public (State, county, and municipality) ownership and a decentralized organizational structure. This structure has largely been maintained after the reform, but some mergers on the generating and distribution side have been seen.

Broadly speaking the same applies to Finland, Sweden and Denmark. However, before deregulation the generating side in these countries was less decentralized and had a larger share of private ownership than in Norway.

5.8.2 Organizational and technical factors

Norway has traditionally had a very decentralized organizational structure. Before the restructuring there were about 70 power generation companies of which the state-owned Statkraft had about 35% of the capacity. The number of distribution companies was about 200. This large number of market participants is of course an advantage in obtaining efficient competition. However the number of independent companies, especially in the distribution sector, was probably too high to be cost efficient. During the last few years there has been some integration (horizontal integration) which has led to cost savings.

Prior to the deregulation, there was a power pool in Norway based on bidding from generating companies. It had been in operation since 1971 and was an inter-utility market basically for surplus production. The price in this exchange market had no impact on the overall price level, and the influence of this market on companies' profit was marginal, see Figure 5.8.

The spot market and the system of price settlement established after the deregulation were to a great extent based on this already existing pool. The price clearing algorithm (SAPRI) was taken over by Nord Pool and used by Nord Pool Spot until 2009.

Some 95% of the electricity generation in Norway is based on hydropower. The Norwegian power system is therefore marked by some features characterized for hydropower:

The system has high investment costs and very low variable costs. Consequently it has low cost variability. This does not necessarily lead to low price variability. Limitations in reservoir capacities and variations in precipitation may cause price variations between seasons and/or between day and night. However, with the reservoir capacity presently available in Norway, reservoir limitations are not causing any large hourly or daily price fluctuations under normal circumstances. Under normal precipitation there is some systematic price variation over the seasons in Norway. However, in very dry years there is a high price level during the winter while a very wet year gives extremely low prices in the summer.

5.8.3 Organizational structure before restructuring

The electricity supply industry in Norway had a three level structure prior to the Energy Act from 1990 as illustrated in Figure 5.4.

Before restructuring, Statkraft was responsible for 1/3 of the generating capacity and 70% of the transmission grid. Statkraft was also responsible for and had a legal monopoly on import/export of electricity. It supplied power consuming industry and regional wholesale utilities with electricity on long term contracts. The regional companies bought electricity from Statkraft (the price on these contracts as well as the price in the pool can be seen on Figure 5.8) and had normally some own generation facilities in addition. The local distribution companies - the retailers - bought electricity from "their" whole sale company.

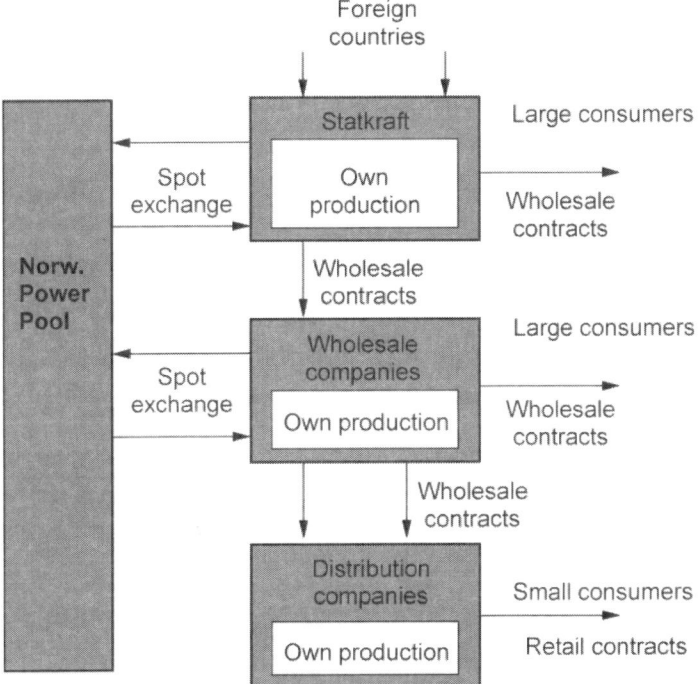

Figure 5.4 Electricity trade in Norway before restructuring.

5.8.4 Major arguments for restructuring

The major arguments for the establishment of a competitive electricity market in Norway can be summarized as follows:

Avoid excessive investment

The pricing policy of the entire electric utility industry was based on cost recovery. Utility companies were able to mix the cost of expensive new development with cheap existing plants in such a way that prices to customers did not reflect marginal costs. Thus it was possible to develop more capacity than consumers were willing to pay for.

Improve selection of investment projects

New supply resource capacity was developed in a sequence that was not optimal for society. Expensive and clearly unprofitable projects were being developed, while less expensive projects were not. This was mainly due to a municipal owner structure, political constraints and the obligation to serve the demand within each local region. It was impossible to rank hydro power projects so that the least expensive ones were developed first.

Create incentives for cost reduction

In a system where prices are based on cost recovery, there is no basic incentive for cost savings. Excessive costs can be passed on to the customer. In a competitive system where the price is settled by the market, cost saving incentives will be created.

Equity among consumers

Prices to customers were in most utilities decided by political institutions in the old regime. The electricity price for the power intensive industry was decided by the parliament, residential prices were decided by municipalities. In many cases this lead to cross-subsidizing. For instance, commercial customers could pay too much and residential customers too little. Power intensive industry generally had a low price. A competitive market would create equity among consuming sectors and improve market efficiency.

Reasonable geographical variations

The location of the natural reasons decides where hydroelectric plants can be built. Vicinity to those sites has justified low prices and has been a basis for the development of power intensive industry in Norway. Ordinary consumers have also derived benefit from vicinity to hydroelectric sites[14]. To some extent the basis for this geographical differentiation disappears with a stronger transmission grid.

5.8.5 The Nordic generating system

There is a different mix of production resources in the Nordic countries.

Table 5.1 gives an overview. Norway has almost 100% hydropower. Sweden and Finland have a mix of hydropower, nuclear and other thermal plants, and Denmark is dominated by thermal production and has a considerable contribution from new renewable sources (wind). The total electricity consumption in the Nordic countries was 400 TWh in the year 2004 and the generation was about 390 TWh.

---

[14] One important mechanism behind this is the so-called *concession power*. Concession to build a hydropower plant in Norway is tied to the condition that the local municipality receives a certain amount of concession power at a low price. This mechanism is still in operation and there is still some geographical variation in Norway after deregulation, although it has been reduced considerably.

Table 5.1 *The Nordic countries, overview 2009. (Source: Statnett).*

		All Nordic	Denmark	Finland	Norway	Sweden
El. consumption	(TWh)	377.5	34.7	80.8	123.7	138.3
El generation	(TWh)	369.7	34.4	68.7	132.8	133.7
Hydro	(TWh)	206.1	0	12.6	128.3	65.2
Nuclear	(TWh)	72.6	0	22.6	0	50.0
Thermal	(TWh)	80.4	27.7	33.3	3.5	15.9
Wind	(TWh)	10.5	6.7	0.3	1.0	2.5

5.8.6 The organization of the new market

There are a large number of participants in the Nordic market. Generating companies and large end users trade in the wholesale market on a bilateral or spot basis. Still the vast majority of small customers have access to the market either through their local distribution company or through an external supplier. Most of the producers are also distributors, and almost all the distributors are traders, that is, they buy power from the wholesale market and sell it both on the wholesale and retail levels.

During the last 10 years, there has been some integration in the Nordic power supply industry. There has been horizontal integration on the generating as well as the distribution levels. The large generating companies have bought smaller ones and increased their market share. There have been several mergers among distribution companies. In some cases we have also seen vertical integration in the sense that generating companies have bought distribution companies. But there are still a large number of market participants which is good for competition.

All participants have access to the power grids, both for transmission and distribution. Transmission is undertaken by the national grid companies: Statnett in Norway, Fingrid in Finland, Svenska Krafnät in Sweden and Eltra and Ekraft (which have recently merged into Energinet) in Denmark. The grid companies are also System Operators. Distribution is undertaken by the distribution companies, which in most cases are owned by counties and municipalities.

The market structure is shown in Figure 5.5 and some additional information is given in Table 5.2 below.

Chapter 5: Restructuring/Deregulation of the Electricity Supply Industry

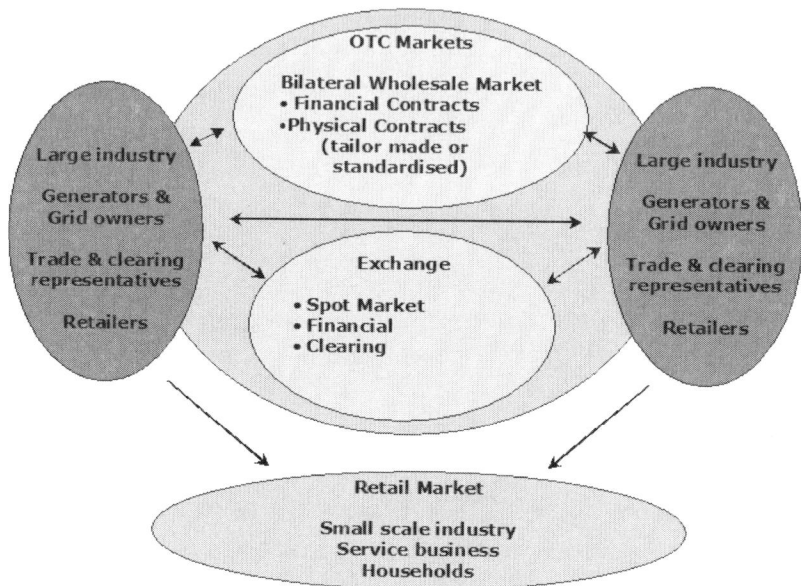

Figure 5.5 Overview of the Nordic electricity market.

Table 5.2 Markets in the Nordic system.

Market place	Physical trade	Financial trade
Nord Pool Spot	Elspot Elbas	
NASDAQ OMX Commodities		Futures Forwards Options CfDs
System Operator	Regulating power market (RPM) RP options (Norway)	
Bilateral (Brokers, traders)	Full delivery Load factor contracts	Forwards Options

On the wholesale level (upper part of Figure 5.5) the market is divided between the so-called OTC market[15] and the Exchange presently comprising Nord Pool Spot and NASDAQ OMX. About 70 % of the physical trade goes through Nord Pool Spot and the rest through the OTC market.

5.8.7 The Exchange - Nord Pool Spot and NASDAQ OMX

Development of the Nordic exchange.

Nord Pool was established in 1993 as an exchange for the Norwegian electricity market. In 1996 the exchange was extended to include Sweden, and thus became the world's first multinational exchange for electricity.

Presently, Nord Pool Spot runs the largest market for electrical energy in the world, offering both day-ahead and intraday markets to its participants. Nearly 340 companies from 18 countries trade on the exchange. The Nord Pool Spot group has offices in Oslo, Helsinki, Stockholm, Copenhagen, Tallinn and London. In 2010 the group had a turnover of 307 TWh representing a value of EUR 18 billion and 74% of the total consumption of electricity in the Nordic countries.

The Nord Pool Spot group is owned by the Nordic transmission system operators.

Nord Pool Spot provides a market place to producers, energy companies and large consumers on which they can buy or sell electrical energy. Nord Pool Spot is the central counter party in all trades, guaranteeing settlement for trade.

Nord Pool Spot's system price is the reference price for futures, forwards and options traded in the financial market. The system price is also the reference price for the Nordic OTC/bilateral wholesale market and used by distributors as basis for quoting prices to end consumers.

The Norwegian Water Resources and Energy Directorate (NVE) regulates Nord Pool Spot and issues the market place concession. The Norwegian Ministry of Petroleum and Energy (OED) has authorized Nord Pool Spot to organize the exchange of electrical energy with neighboring countries.

Nord Pool established a forward market as early as 1993. That was a physical market, but it later changed to a financial arrangement. From 2009 NASDAQ OMX Commodities took over that activity and has expanded to include derivatives related to the German (EEX), Dutch (APX) and the UK (N2EX) spot market.

---

[15] OTC is short for Over The Counter. It includes different types of contracts.

Products and services

Elspot is the spot market offering trade for day-ahead physical delivery. Prices are determined through a double auction (auction with supply and demand participation) for each hour in the day. The System Price, which is the unconstrained price in Elspot (based on the assumption that there are no transmission restrictions in the grid), is the reference price for financial trade in the Nordic market.

Norway is subdivided into two or more geographical price areas or bidding areas. These areas are fixed for a period for at least 3-4 months. Sweden, West- and East-Denmark, and Finland constitute one separate price area each. From November 1 2011 Sweden is divided into 4 bidding areas. In situations where capacity limitations in the main grid occur, the day-ahead market is used to relieve congestions between the price areas. After the System Price has been settled, Area Prices are determined in such a way that lines otherwise overloaded will stay within capacity limits. A detailed description is given in Section 7.2.

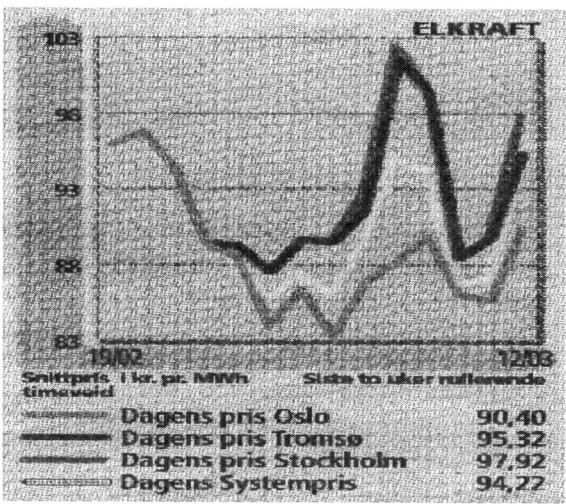

Figure 5.6 Spot prices (System Price and Area Prices) published in a Norwegian newspaper (Dagens Næringsliv).

Elbas is a short-term market for physical power trading. It enables trading close to real time around the clock every day of the year, covering individual hours up to one hour before delivery. Its function is to be the aftermarket to the Spot Market at Nord Pool Spot, giving power market participants the opportunity to adjust their balance closer to real time. The participants are producers, distributors, industries and brokers. The Elbas market was for some time limited to Finland and Sweden only, but now the whole Nordic area is included.

Futures and Forwards are products in NASDAQ OMX's financial market. They are derivatives, which means they derive from an underlying market, i.e. the spot market. Futures and Forwards are instruments that are primarily used for price hedging and risk management related to trade in electric power. The Futures have a trading horizon of 8 – 9 weeks whereas Forward contracts have a horizon of maximum 6 years. Both are financial contracts with no physical delivery, only a financial settlement against the System Price. Prices in Elspot (System Price), Futures and Forwards are shown in Figure 5.7.

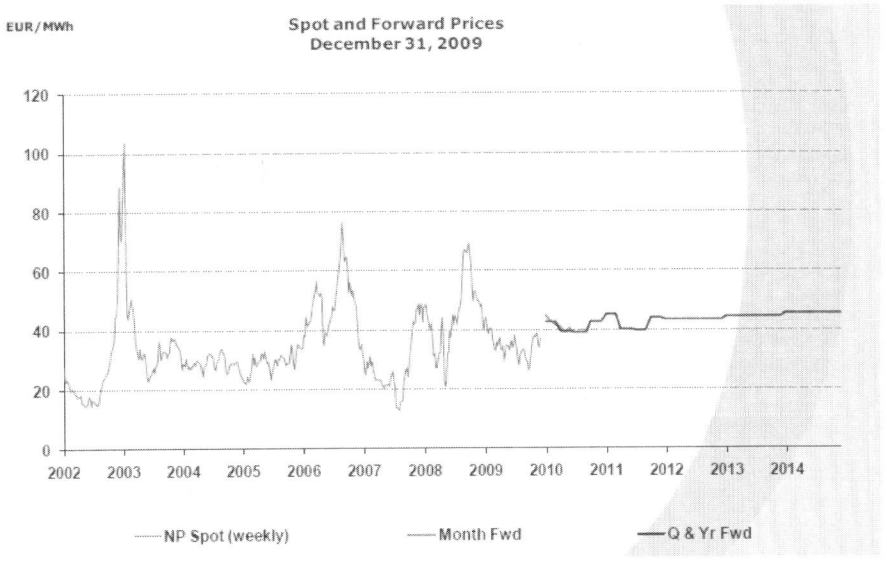

Figure 5.7 Prices at the Nordic exchange.

Options are financial derivatives also used for price hedging/risk management related to trade in electric power. European-type[16] power options with Forward contracts as underlying instruments are offered at Nord Pool.

Contracts for differencses (CfDs) can be purchased to hedge spatial price differences between price areas. It is also possible to hedge the price area risk by trading a set of CfDs for the difference between the system price and the different area prices.

---

[16] There are different types of options: European, Asian, etc. Concerning details on options and other financial instruments, see [7]

Clearing of power contracts through NASDAQ OMX Commodities clearinghouse reduces financial counter party risk for market participants. The clearinghouse enters into financial power contracts and carbon contracts as a contractual counterparty. The clearinghouse assumes liability for covering the future settlement of these contracts, thereby reducing the risk for both buyer and seller. NASDAQ OMX Commodities provides clearing for standardised contracts traded on the exchange and in the over-the-counter (OTC) market.

5.9 Developments in the Norwegian and Nordic market

5.9.1 The pool's influence on the general price level

The operation of the Norwegian power system has traditionally been based on decentralized decisions. There has never been a central operation scheduler or dispatcher. There was central monitoring and coordination for which the former power pool (Samkjøringen) was responsible. However, there was never any central optimization of operations. The operational decisions were taken by each individual generating company.

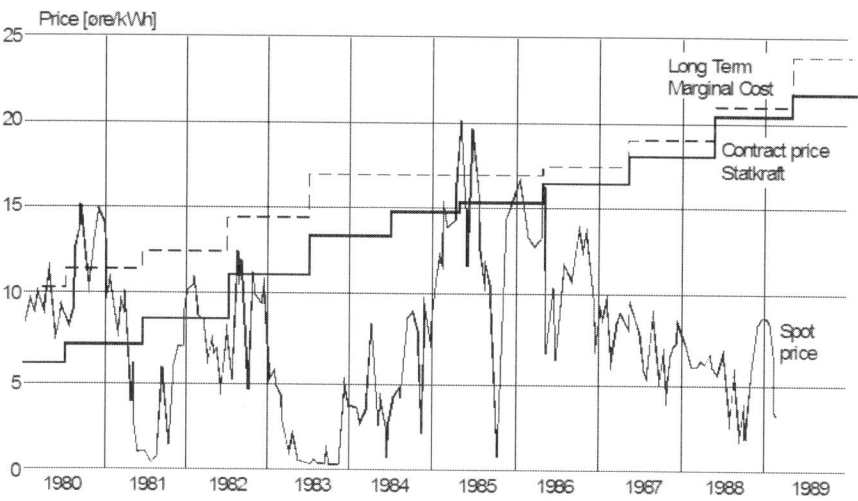

Figure 5.8 Spot and wholesale contract prices before deregulation.

Since 1971 a short-term power exchange market has been in operation under the administration of Samkjøringen. Until the restructuring in 1991 this spot market was an inter-utility market with certain basic restrictions on participation:

- End users were not allowed to enter[17].

- Utilities could only participate if they had an acceptable power balance, i.e. all firm power obligations had to be covered by their own generating capacity or by long-term contract purchases.

- The power balances of participating utilities were controlled by Samkjøringen once a year through power balance calculations. These calculations included annual energy as well as peak capacity.

As a consequence of these restrictions on participation, the prices in the spot market had no influence on the long-term contract prices or on the tariffs. This is illustrated in Figure 5.8 showing the development of the spot price, the Statkraft contract price and an estimated Long Run Marginal Cost (LRMC).

From this figure we can see:

- There was no connection between the spot price and the contract price.

- The contract price was close to LRMC reflecting the official pricing policy at the time. It can also be seen from the figure that there was a certain gap between the two, reflecting the fact that this official pricing policy was never fully implemented (not even for the State-owned utility Statkraft).

- On average the LRMC (and the contract price) was considerably higher than the Short Run Marginal Cost (SRMC) represented by the spot price. This indicates that the production system was oversized. In 1990 it was estimated an expected SRMC of 12 – 13 øre/kWh compared to an LRMC of approximately 26 øre/kWh. The corresponding overcapacity was estimated to 6-7 TWh/year.[18] (See Section 4.2)

The price settlement system in this spot market (24-hour market) was initially kept almost unchanged after the restructuring. The most important changes were that the market was opened to participation from end users and all restrictions and controls on the participating utilities were removed. These changes decisively increased the influence of the spot market on the general price level. This is clearly shown in Figure 5.9.

---

[17] Certain types of end use could be covered by this spot market. The most typical example is dual fuelled boilers based on electricity and oil.

[18] It should be emphasized that the Norwegian system was (and still is, although not to the same extent) an <u>energy constrained system</u> meaning that the amount of water available in the most critical (i.e. the driest) year is decisive. This makes it easy to draw this type of conclusions from these simple observations.

Figure 5.9 Spot and wholesale contract prices after deregulation (Source: Norsk Kraftmegling).

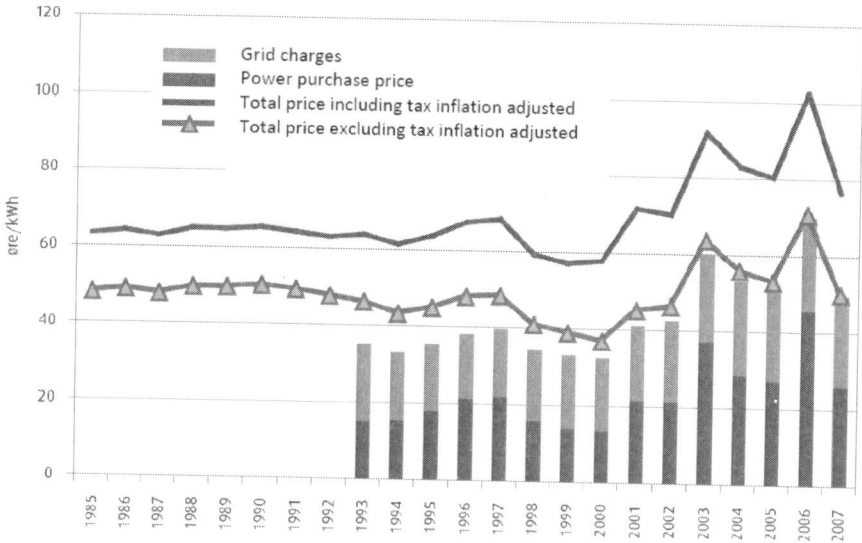

Figure 5.10 Household prices, transmission/distribution included (Source: OED).

However, Figure 5.9 shows wholesale prices. The price paid by the end user will not necessarily follow the same pattern. The experience from the first years after deregulation was that end user tariffs were only affected to a small extent by the wholesale prices, and that the effect was delayed. The development in household prices is shown in Figure 5.10. The same is evident from Figure 5.11.

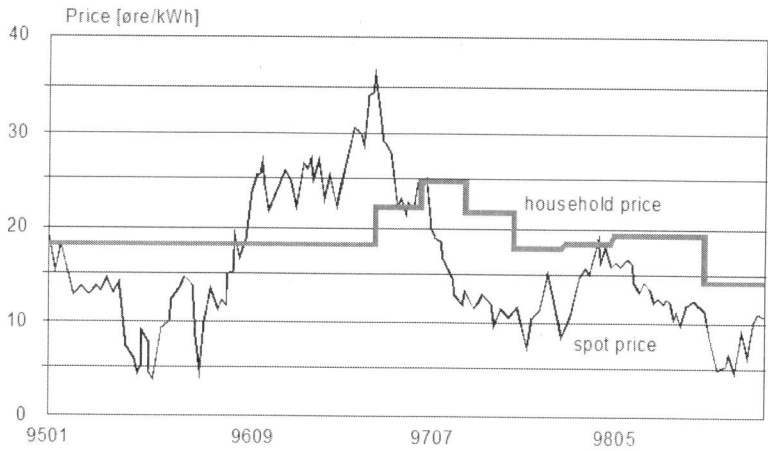

Figure 5.11 Spot and household price 1995 – 1998.

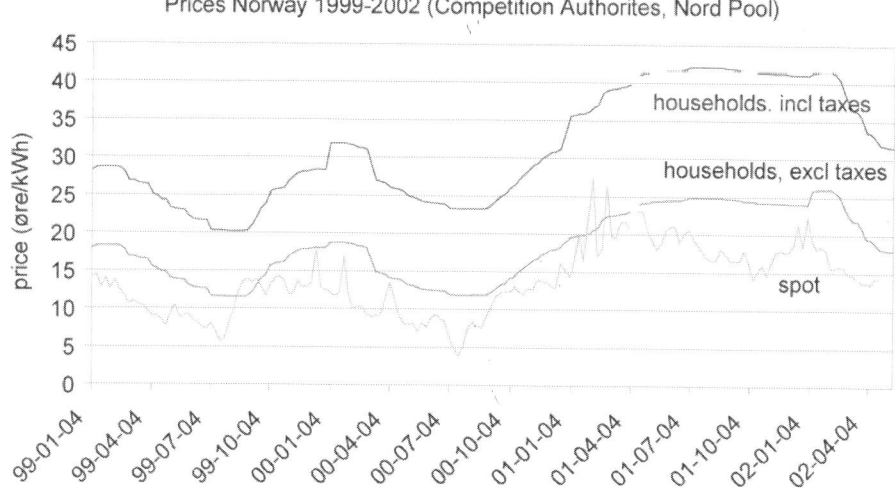

Figure 5.12 Spot and household prices 1999 – 2002.

During the last few years the time lag has decreased as shown in Figure 5.12. The reason for this is primarily more frequent meter reading and simplified procedures for a change of supplier.

Price variability between groups of consumers and regions has decreased after restructuring. This is in line with the argument for the new system. It is a strong indication that competition affects consumer prices.

5.9.2 Investments and power balance

As already mentioned earlier in this chapter, the production system was oversized prior to 1991. Together with unusually high precipitation in the first few years, the overcapacity led to low prices and the willingness to invest in new capacity dropped to zero. A low rate of investment and a continued growth in demand have led to serious consequences for the national supply balance.

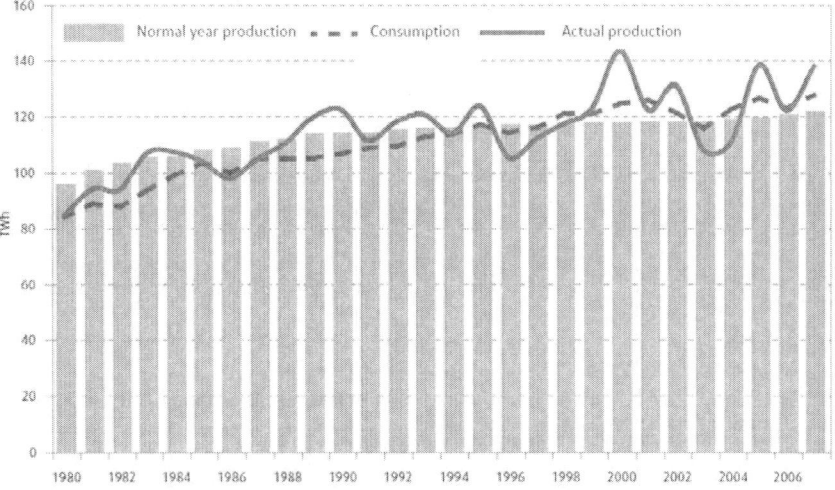

Figure 5.13 Energy balance in the Norwegian system.

Figure 5.13 shows the development of the Norwegian energy balance. From having an energy surplus in a normal year, Norway has developed an energy import dependency. There is still a surplus of capacity (see Figure 5.14), but it is narrowing. In the course of a few years Norway will probably have to depend on peak capacity import.

In the longer run, prices are expected to increase and investment will presumably pick up again. But there is still some uncertainty regarding future investments in the deregulated power generating system [6].

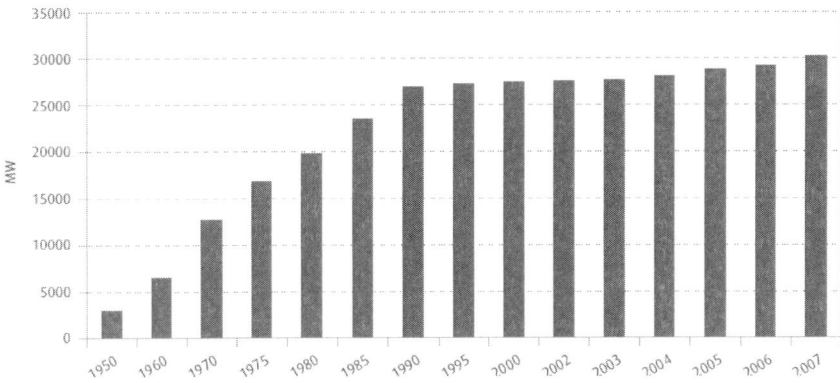

Figure 5.14 Development of generation capacity in the Norwegian system.

5.9.3 Market Efficiency

Market efficiency seems to be reasonably good in the Norwegian (and later the Nordic) market. The spot price development gives an indication of this. Certain incidents such as the price peak in February 1994 (see Figure 5.9) can be seen as an indication of some price manipulation (market power). However, it has not been found necessary for the regulatory authorities to interfere with price control measures.

In order to study market efficiency SINTEF Energy Research has performed central scheduling studies based on the EMPS model [2], and compared the simulated results with what actually has happened in market place. Figure 5.15 shows a pre-deregulation study. The correlation between simulated prices, that reflect current Short Run Marginal Cost (SRMC), and the prices observed in the spot market, is quite good.[19]

---

[19] The fact that prices are close to SRMC is generally an indication that the market is efficient. That is the case for the Norwegian system before restructuring too, except for the obvious fact that most of the consuming side had no access to this short-term variable price. As pointed out earlier, Samkjøringen's spot market was only accessible for generators and the limited number of consumers that were defined as flexible, for instance dual fuel (oil and electricity) boilers. The market was efficient on the generating side: The generators adapted to SRMC. But it was inefficient on the consuming side. The consumers were not confronted with the SRMC.

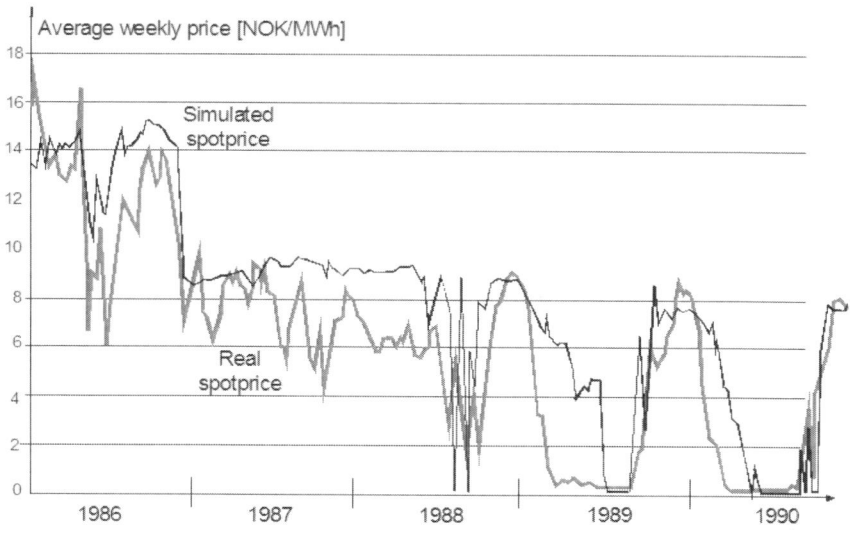

Figure 5.15 Simulated and observed spot prices, pre-deregulation.

Figure 5.16 shows a similar study done after the deregulation. The general impression is that the correlation between the simulated and real prices continues to be good, which indicates a high market efficiency.

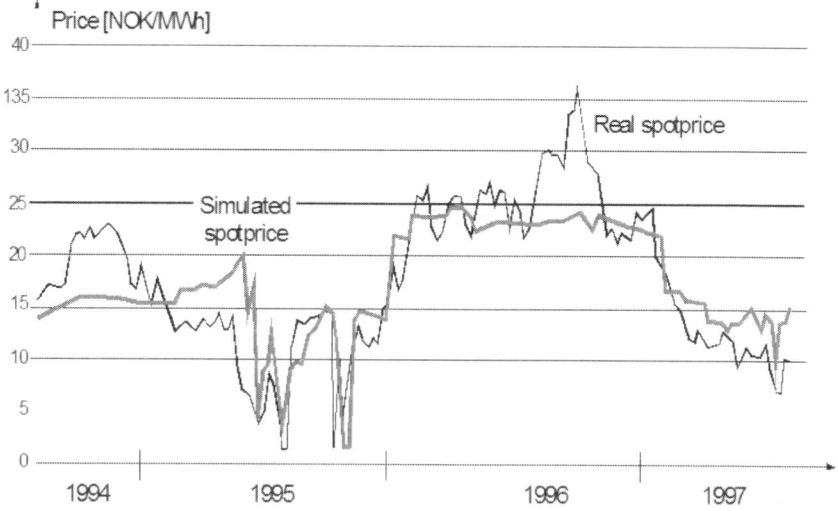

Figure 5.16 Simulated and observed spot prices after deregulation [2].

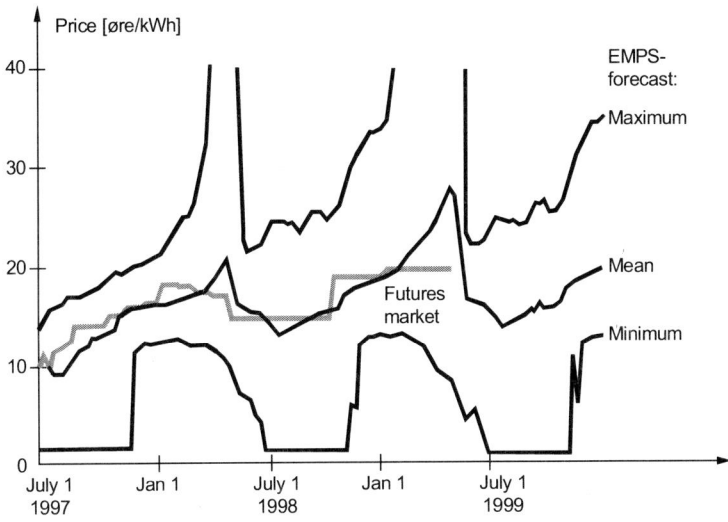

Figure 5.17 Price forecast based on the EMPS model and prices from the Futures marked [3]

Figure 5.17 shows the result of spot price forecasting done with the EMPS model and the corresponding prices in the Futures market. The same general conclusion can be drawn: there is a close correlation between the prices from the model and observed prices in the market, indicating an efficient market.

5.9.4 Market growth at Nord Pool

Figure 5.18 gives an overview of the development in the Nord Pool market.

The turnover in Elspot has grown steadily over the last decade. Notice that the numbers are in GWh/week. The turnover in 2004 was about 40% the total Nordic generation.

The financial products (Futures, Forwards, options and CfDs) showed a growth in turnover until 2002. Futures and Forwards represent the bulk of this market. The liquidity in the option and CfD market is low. The development of clearing services has been similar to the financial products. Notice that numbers in both these graphs are in TWh/week. Total turnover in these two markets are 2 – 3 times the total Nordic generation. That means that much of the electricity is traded several times before it is physically delivered.

5.9.5 Market information and statistics

Nord Pool is the leading market data provider delivering information from the Nordic power market. The information system has been one of the important criteria for the success for the power exchange. Nord Pool's website distributes statistics concerning production, consumption, exchange, hydropower reservoirs etc. Urgent market messages (UMM) concerning maintenance and failures are updated within short time limits. In this way all market participants get equal information at the same time. The companies holding UMM-information are defined as insiders until the information is published, and are not allowed to take financial positions on the exchange in this period.

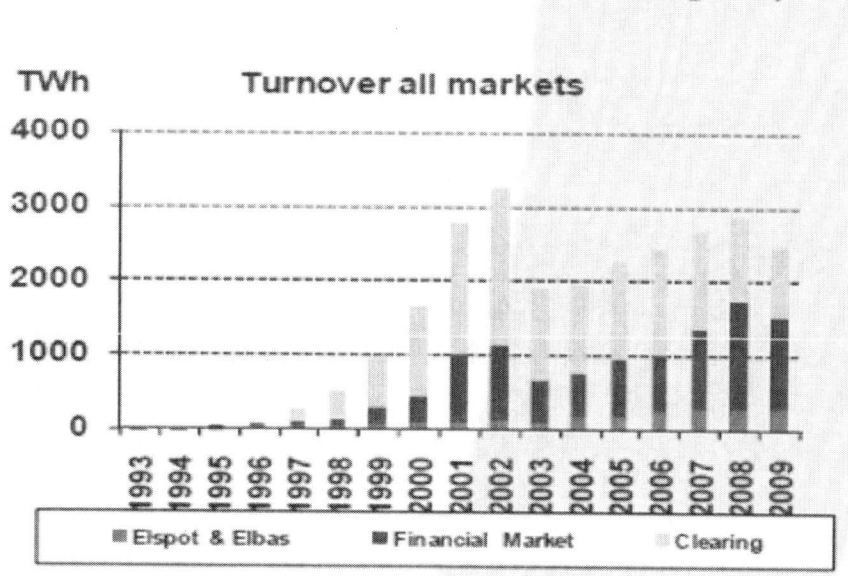

Figure 5.18 Traded volumes on Nord Pool. (Source Nord Pool).

5.10 Retail access

All the Nordic countries introduced in principle full retail access right from the start. The exception was Denmark where full retail access was not implemented until 2003.

However, at the beginning there was a requirement for hourly metering and that had to be paid by the consumer. It was therefore unattractive for small consumers to switch suppliers. Gradually the requirement for hourly metering has been replaced by a profile-based settlement for small consumers. Profile-based settlement implies that each consumer's hourly demand is estimated based on a chosen profile. Slightly different rules are applied. In Norway there is one profile for each distribution company (total demand profile less all hourly metered consumption) is used, whereas Finland and Sweden use different profiles for different classes of consumers.

A fee for change of supplier was used at the beginning. That was gradually decreased and completely removed in Norway in 1997. Change of supplier is now a simple process, mainly involving conveying the necessary information to the new supplier. Price information is readily provided on the Internet by the competition authorities, and a number of suppliers actively advertise in the media. The number of consumers switching to external suppliers has increased significantly in Norway since 1998. But still more than 70% are using the local supplier. See Figure 5.19. We notice that there is high activity in the market in periods of high and rapidly changing prices, for instance 2002/2003.

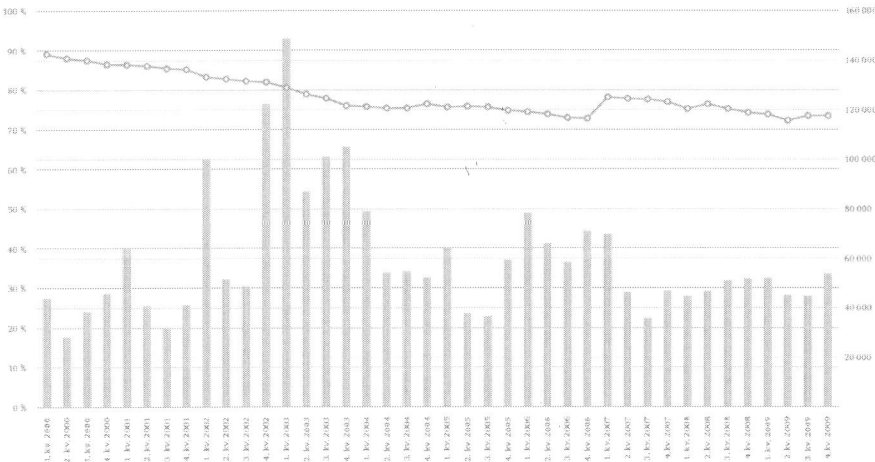

Figure 5.19 Number of household customers changing supplier every quarter (bars) and percentage of customers sticking to their local supplier (line).

In Sweden, the share of consumers buying from external suppliers is presently about half of what we have in Norway. This difference is most likely caused by the fact that Norwegian consumers have had the opportunity to switch between suppliers for a longer time than Swedish customers.

Also Finland has used the same principles from November 1998. However, the number of small consumers that have changed supplier is considerably lower than in Sweden and Norway, presently 3-4 per cent. For commercial and industrial consumers, the numbers are similar to Norway.

5.11 References

[1] S. Aam, I.Wangensteen: "Deregulation of the Norwegian electricity supply industry" World Energy Counsil, Houston, Texas 1998

[2] M. Brennvik, A. Foshaug: "Langsiktige prisbevegelser i kraftmarkedet". Hovedoppgave NTNU, 1997.

[3] O.B. Fosso et.al. "Generation scheduling in a deregulated system. The Norwegian case", IEEE Winter Meeting 1998

[4] N. Flatabø, G. Doorman, O.S. Grande, H. Randen, I Wangensteen : "Experience with the Nord Pool Design and Implementation." IEEE

[5] R. Wilson: "ARCHITECTURE OF POWER MARKETS" Econometrica, July, 2002

[6] A. Botterud: "Long-Term Planning in Restructured Power Systems" PhD thesis Department of Electrical Power Engineering NTNU, 1993.

[7] Nord Pool: "Trade at Nord Pool's Financial Market" March 2006

[8] Nord Pool: "Trade at the Nordic Spot Market" March 2006.

[9] O. Wolfgang, A. Haugstad, B. Mo, A. Gjeldsvik, I. Wangensteen, G. Doorman: "Hydro reservoir handling in Norway before and after deregulation." Energy 34 (2009) 1642 – 1651

[10] Xiao-Ping Zhang: "Restructured Electric Power Systems"
John Wiley & Sons 2010

[11] Sally Hunt: "Making competition work in electricity"
John Wiley & Sons 2002

[12] http://www.pjm.com/

[13] Yadav. Roy, Khaparede, Pentaya: "India's Fast-Growing Power Sector" IEEE power & energy magazine, July/August 2005

[14] http://www.npspot.com/

References

[15] I. Androcec, S. Krajcar, I. Wangensteen: "Optimization of costs and benefits in Inter TSO Compensation mechanism" EEM 11, Zagreb, 25 – 27 may, 2011

6 Power generation

6.1 Introduction

This chapter looks at the power generating system. We discuss simple methods for operation planning for thermal and hydropower units. On the investment side, it is shown how a simple technique, a so-called *screening technique*, can be used to find an optimal mix of different thermal units.

The Chapter starts with operation planning under conditions that are largely in line with the situation before deregulation. It is assumed that the utility is confronted with a price-quantity relationship on the consumption side and the utility has maximum social surplus as its objective. It is explained how models for this type can be used as market simulation models.

This planning situation is basically different from a competitive environment, where the generating company has profit maximization as objective and operates as *price taker* in the market. Principles for this planning under such conditions are described in Section 6.5 and practical procedures in Section 6.6

The power grid is disregarded in this chapter. In Chapter 7 it is shown how the grid can be included so we extend the *Optimal Dispatch* described here to include the grid in so-called *Optimal Power Flow*. It is also shown how transmission pricing can have impact on hydro power scheduling.

6.2 Characteristics of generating units

6.2.1 Thermal units

The cost for a thermal unit consists, in line with general cost description in Chapter 2, of a fixed (operation independent) and a variable (operation dependent) cost:

$$C = b + \Phi(P) \qquad (6.1)$$

P is output in MW, b is a constant.

The fixed cost is dominated by capital cost (depreciation and interest) and some fixed operation and maintenance costs. These costs accrue regardless of operation.

The operational cost is dominated by fuel costs. Thermal efficiency and fuel price are decisive factors. Cost as a function of output can be as shown in. Figure 6.1

There is a minimum and a maximum output and there is an operation cost which rises with increasing output.

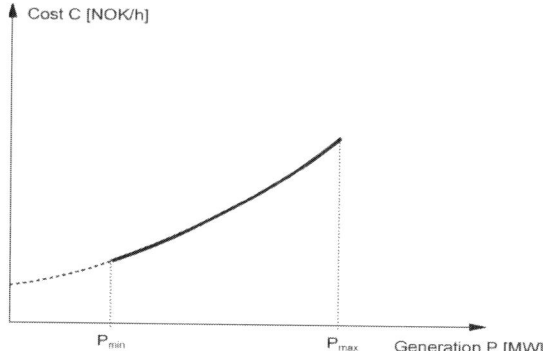

Figure 6.1 Operation cost for a thermal unit.

The operation cost for a thermal unit is often represented by a quadratic function:

$$\Phi(P) = \alpha + \beta P + \gamma P^2 \qquad (6.2)$$

where α, β and γ are constants. The constant α represents the fixed cost of keeping the unit running (with zero output). A thermal unit requires ancillary power in order to be in operation in addition to staffing, and the cost is largely independent of output. The fixed (operation independent) cost and the fixed part of the operation cost do not affect the marginal costs which can be expressed as:

$$c = \frac{dC}{dP} = \beta + 2\gamma P \qquad (6.3)$$

The marginal cost of a thermal unit will be a linearly increasing function of output (provided $\gamma > 0$) as shown in Figure 6.2.[1] As an approximation in some cases we regard the marginal cost as constant. The optimization of a set of units will then be very simple (merit order loading).

---

[1] In power system analysis it is often assumed that the marginal cost is increasing with rising output. See [1]. In most practical cases, however, the marginal cost is decreasing with rising output.

Characteristics of generating units

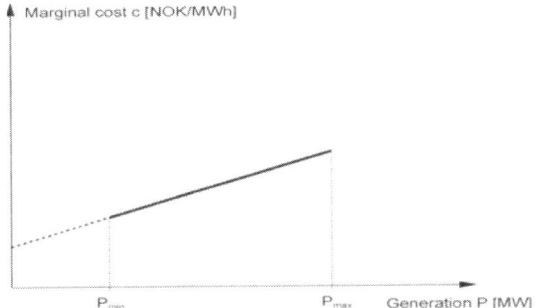

Figure 6.2 Marginal cost for a thermal unit.

In a dynamic context we have to take additional costs and restrictions into account:

- Start up costs
- Restrictions on minimum uptime and minimum downtime.
- Ramping restrictions.

These costs and restrictions are more important for thermal than for hydro units.

6.2.2 Hydropower units

Output, P, for a hydropower unit depends on the head [m], the flow of water [m³/s] and the efficiency according to the following formula:

$$P = \rho \eta g Q H \qquad (6.4)$$

where:
H is the head in m
Q is the water flow in m^3/s
g is the gravitational acceleration m/s^2
ρ is the density of water kg/m^3
η is the efficiency

The output is proportional to the head, which in principle is affected by the release of water. Emptying a reservoir will reduce the head. That is taken into account in some models for operation planning, but for simplicity reasons it is disregarded here. We regard the head as constant.

The output is proportional to the release of water if all other factors are constant.

The efficiency is slightly variable. A typical curve is shown Figure 6.3. Notice that the efficiency is generally high and that maximum efficiency is reached at an output level below maximum.

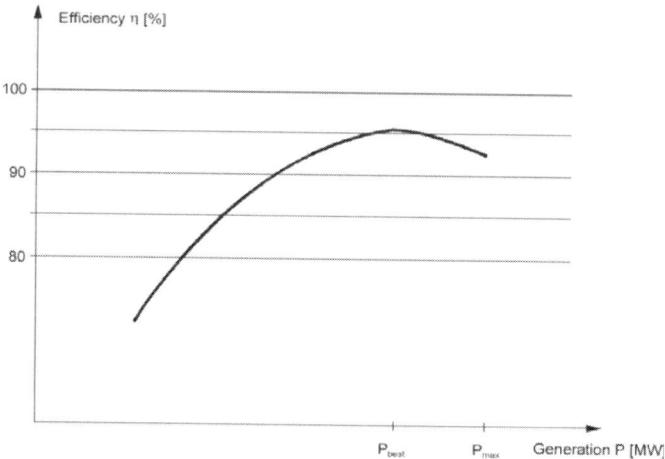

Figure 6.3. Efficiency curve for a hydroelectric unit.

A hydropower plant with several units will have decreasing overall efficiency as more units are started. Figure 6.4 shows measured efficiency for a Norwegian power plant with four units. The loss in the waterway (penstock etc.) rises with an increasing amount of water.

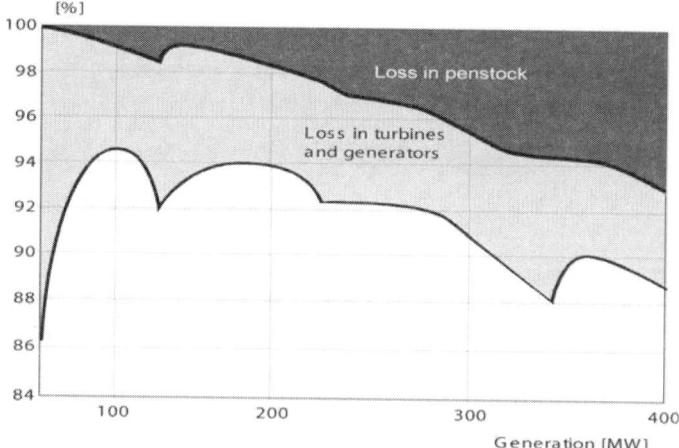

Figure 6.4 Efficiency for a Norwegian four unit hydropower plant.

Generally, the efficiency for a modern hydropower plant is high. There is some variation depending on output and detailed optimization models take this variation into account. In the simple models we discuss here, we disregard this variation. We assume output is proportional to the release of water:

$$P = kQ \qquad (6.5)$$

where k is a constant. It is then possible to quantify an amount of water in kilowatt-hours. We follow that in the examples below.

Figure 6.5 shows simple one reservoir model of a hydropower plant which in many cases can be more complex. The flow of water to a hydro plant can be divided between so-called *controllable inflow* into the reservoir *uncontrollable inflow* outside. There is a generating station with an installed capacity. Water can be lost through overflow or bypass. Planning of hydro generation is always a question of how to dispose of the reservoir. It is a dynamic planning problem.

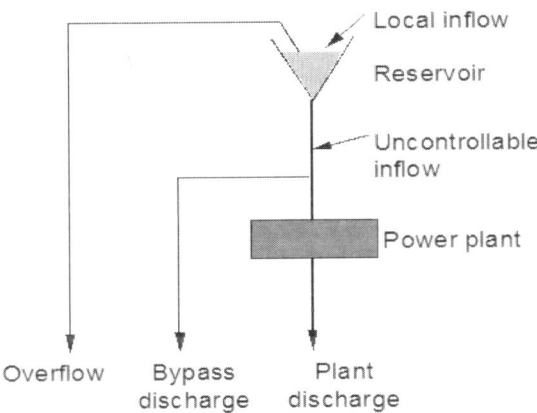

Figure 6.5 The one reservoir model.

6.3 Generation scheduling

The complexity of generation scheduling depends a lot on intertemporal ties, i.e. to what extent it is taken into account that a decision at one point in time affects the situation and has impact on decisions at another instant of time. In a hydropower system with reservoirs, intertemporal ties will always have to be included. In a thermal system we can use simplified procedures without any regard to the time dependency.

6.3.1 Merit order

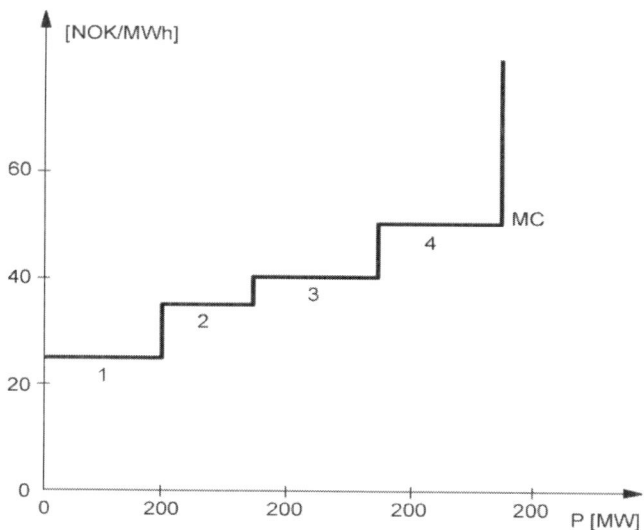

Figure 6.6   Merit order loading. Four units with increasing marginal cost.

Merit order loading means the units are started with the cheapest first and then we proceed in merit order with respect to cost as the load increases. Merit order is based on constant marginal cost for each unit. It is a simple procedure. The simplification is not only that the marginal cost is regarded as constant, but start-up costs and restrictions mentioned above are disregarded. That makes it possible to do the operation planning without taking intertemporal ties into account.

6.3.2 Optimal Dispatch

If the marginal cost is variable depending on the output, we are faced with the problem of dividing total load between available units in an optimal way. In this case optimal means minimum cost. This is a traditional optimization problem.

We look at a case with two units. The objective is to minimize the total cost:

$$Min[C(P)] = Min[C_1(P_1) + C_2(P_2)] \qquad (6.6)$$

subject to the restriction:

$$P = P_1 + P_2 \qquad (6.7)$$

Generation scheduling

where P is the total load to be covered.

We solve it in the traditional way by introducing a Lagrange multiplier and defining an extended objective function (Lagrange function):

$$L(P_1, P_2, \lambda) = C_1(P_1) + P_2(P_2) + \lambda(P - P_1 - P_2) \quad (6.8)$$

The necessary conditions for minimum cost are:

$$\frac{\partial L}{\partial P_1} = \frac{\partial C_1}{\partial P_1} - \lambda = 0 \quad (6.9)$$

$$\frac{\partial L}{\partial P_2} = \frac{\partial C_2}{\partial P_2} - \lambda = 0 \quad (6.10)$$

$$\frac{\partial L}{\partial \lambda} = (P - P_1 - P_2) = 0 \quad (6.11)$$

The result (as we know from economic theory, see Chapter 2) is that marginal cost is to be the same for both units. In power system analysis we call this *system lambda*: the marginal cost which is equal for all units in operation.

$$\frac{\partial C_1}{\partial P_1} = \frac{\partial C_2}{\partial P_2} = \lambda \quad (6.12)$$

A simple example with second-degree cost functions:

$$C_1(P_1) = 2400 + 30P_1 + 0.06P_1^2 \qquad (6.13)$$

$$C_2(P_2) = 3600 + 24P_2 + 0.09P_2^2 \qquad (6.14)$$

and the restriction:

$$P_1 + P_2 = 900 MW \qquad (6.15)$$

Equation (6.13) and (6.14) yield:

$$\frac{\partial C_1}{\partial P_1} = 30 + 0.12P_1 = \lambda \qquad (6.16)$$

$$\frac{\partial C_2}{\partial P_2} = 24 + 0.18P_2 = \lambda \qquad (6.17)$$

The set of equations:(6.15),(6.16) and (6.17) can easily be solved with respect to the unknowns: λ, P_1 and P_2:

λ = 92.4 NOK/MWh
P_1 = 520 MW
P_2 = 380 MW

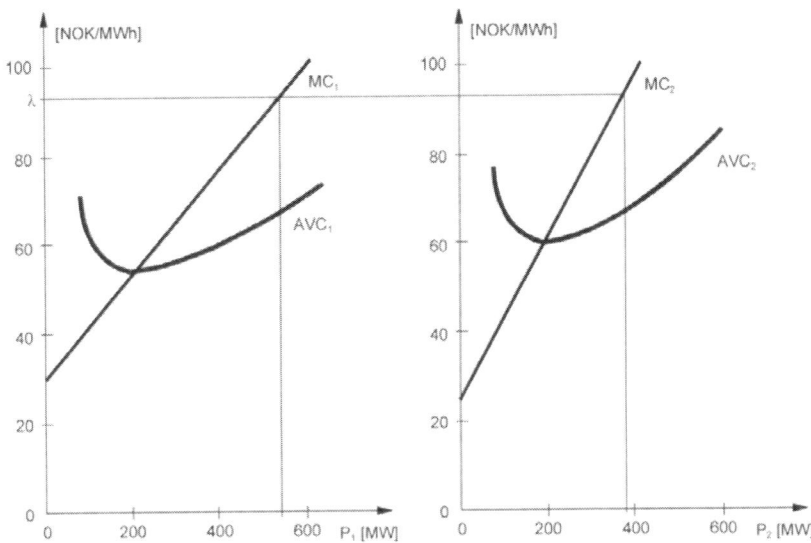

Figure 6.7 Marginal cost (MC) and average variable cost (AVC) for two units in operation. Optimal dispatch is represented by equal marginal cost, system lambda.

The optimality condition expressed in Equation (6.12) is based on the assumption that both units are in operation. That is not necessarily the case. In addition to the dispatch problem we therefore have to address a *unit commitment* problem.

6.3.3 Unit commitment

The unit commitment problem is the problem of deciding which units should be in operation and which units should be out.[2] In order to investigate this, we look more closely at the numerical case above. We now leave the assumption that both units are in operation and compare a situation where both units are in operation with a situation where only one unit (in this case unit 1 which has the lowest AVC for a given output) is producing. Figure 6.8 shows the cost characteristics for this two situations: MC and AVC for unit one and MC and AVC unit one plus two. We assume the two units are optimally dispatched as output expands.

---

[2] Units not in operation can be in a standby or reserve mode or they can simply be out of operation. We will not distinguish between different non operational states here. We return to this in more detail in Chapter 9 on Ancillary services.

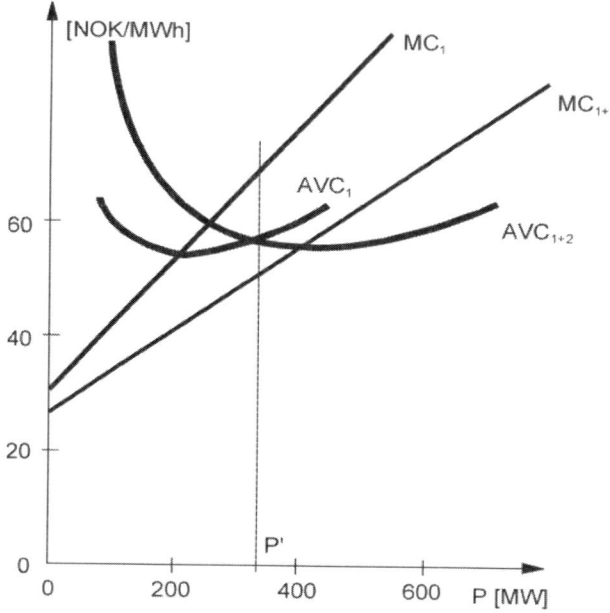

Figure 6.8 Cost characteristics for one unit in operation compared with two optimally dispatched units.

From this figure we can see that the AVC is lower for one unit in operation than for two, as long as the total load is below P'. That means that optimal operation of this two unit system as we increase output, is that unit 1 is in operation until we reach P'. Then we start unit 2.

The reason for this is that there is a fixed cost element in the operation cost. That fixed cost affects the AVC. That fixed cost is higher and affects the AVC more with two units in operation than with one.

With only two units, the unit commitment problem can be easily solved. In principle we have four alternatives:

Unit 1	Unit 2
Off	Off
On	Off
Off	On
On	On

With three units available we have eight alternatives. The general rule is that n units give 2^n alternatives, for instance ten units can be combined 1024 alternative ways. Normally we can rule out some of the alternatives as infeasible and in most cases it can easy be seen that some alternatives will be more costly than others. But still there will be many alternatives to investigate.

6.3.4 Scheduling

The problems we have discussed so far can be solved without taking the time dependency into account. The optimization is done independently for one time step at a time.

The scheduling problem includes time dependency. The problem can be broken down in two sub-problems: a dispatch problem and a unit commitment problem. The dispatch problem is basically the same as the one described above, but the unit commitment is now more complex. In this dynamic context we have to take into account all the time dependent restrictions and cost elements of generating units:

- Start-up and shut-down costs
- Restrictions on minimum uptime and downtime
- Ramping restrictions (restrictions on up ramping, i.e. maximum increase of output pr time unit)[3]

Table 6.1 Ramping restrictions for different generating units. P_N is nominal i.e. full output.

Type of unit	Max up ramping	Unit
Oil or gas	8	% P_N/min
Coal	4 – 8	% P_N/min
Nuclear	5 – 10	% P_N/min
Hydropower	1.5 – 2.5	% P_N/sec

The decomposition of the scheduling problem into a dispatch and a unit commitment problem can be done as indicated in Figure 6.9. The dispatch problem can easily be solved as we have seen above. It can be solved for one time step at a time without regard to the intertemporal links. These links have to be taken into account in the unit commitment model. Therefore it has to be solved for the whole planning period. There is an iterative procedure with exchange of information between the two models as indicated in the figure. The dispatch model receives information on what units are in operation and information is returned to the unit commitment model on the operational cost for the optimally dispatched units.

---

[3] In principle down ramping should also be taken into account, but in practice this is not a problem

We will not describe how this scheduling problem can be solved here. Information about that can be found for instance in [2] and [3].

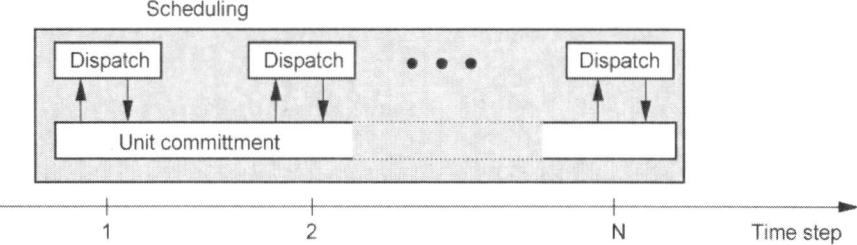

Figure 6.9. Decomposition of the scheduling problem.

6.3.5 Hydro scheduling

One can broadly distinguish between two types of hydro plants: run-of-river plants and reservoir plants. A run-of-river plant has to produce whenever water is available. If we refer to Figure 6.5 all inflow is defined as uncontrollable. Water not used for generation immediately will be lost. Scheduling is therefore irrelevant. Operation of a reservoir plants on the other hand must be planned over a certain time period in order to find an optimal way to dispose of the water.

Simple model for hydro scheduling

In order to explain some basic elements of hydro power scheduling we are using a very simple model.

- We assume one single reservoir and one power plant in line with what is shown in Figure 6.5. We disregard uncontrollable inflow.
- The generation per m^3 of water released through the plant is constant. That means we are disregarding efficiency curves as indicating in figure 6.3 and 6.4 and we are disregarding the impact of variable head which depends on the reservoir content. That means the inflow and release of water can be measured in kWh instead of m^3.
- We regard only two time periods which for instance can be associated with summer and winter. Decisions concerning the use the reservoir are made at the start of each period.
- Inflow Q_1 and Q_2 in the two periods are measured in kWh and the same is the case with release of water (which equals the consumption) W_1 and W_2.
- The generation capacity is sufficient to cover peak load.
- The variable cost is zero.
- The interest rate is zero.

The amount of water is limited and there is limited storage capacity in the reservoir.[4]

The bathtub model

We start with a deterministic situation. Inflows to the reservoir are Q_1 and Q_2 in the two time periods and consumption in the same periods are W_1 and W_2 respectively. We use a so-called bathtub diagram to analyze the case. It is shown in Figure 6.10. Inflow of water in period 1 is indicated by an arrow from left to right starting at the left vertical axis. Correspondingly, inflow of water in period 2 is indicated by an arrow from right to left starting at the right vertical axis. Production (which equals consumption) is treated in the same way. So the distance between the two vertical axes represents the total amount of water available (Q_1 + Q_2) and that equals total production (W_1 + W_2). So there is no net change of reservoir level from the start of period 1 to the end of period 2. Demand in period 1 is indicated by D_1 (increasing consumption as we move from left to right) and the price (marginal willingness to pay) is shown on the left vertical axis. Demand in period 2 is indicated by D_2 (increasing consumption as we move from right to left) and the price (marginal willingness to pay) is shown on the right vertical axis. So it is a conventional demand function in both periods with increasing consumption as the price falls, but for D_2 the diagram goes from right to left instead of the conventional direction. Prices as function of consumption in the two periods are:

$$p_1 = p_1(W_1)$$
$$p_2 = p_2(W_2)$$
(6.18)

The capacity of the reservoir is R. A more detailed discussion of this simple case is found in [10].

---

[4] Compared to practical hydro power planning this is a very simple model. The time resolution in planning models is normally no longer then one week.

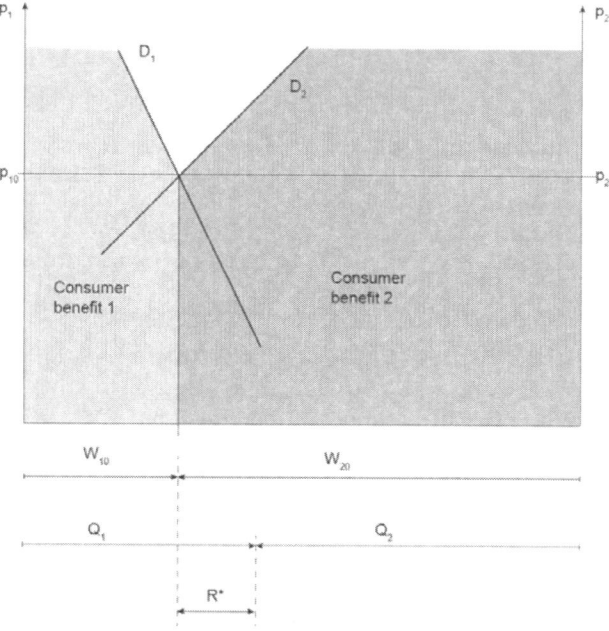

Figure 6.10 Bathtub illustration of a two-period hydro scheduling problem.

The problem is now to decide how much should be produced (and consumed) in period 1 and how much in period 2.

We start with the assumption that the objective is maximum social surplus, which, as long as we have zero variable cost, means maximizing consumer benefit i.e. the sum of area under the demand curves:

$$\text{Max}\left[\int_0^{W_1} p_1 dW + \int_0^{W_2} p_2 dW\right] \qquad (6.19)$$

The consumer benefit in the two periods is indicated by the shaded areas in Figure 6.10. It is obvious that the sum of the two areas is maximized if the price is same in the two periods. That means at the crossing point of the two demand curves. That will lead to maximum consumer benefit. Optimal production in the two periods will then be W_{10} and W_{20}, again referring to Figure 6.10. If we deviate from that (see Figure 6.11 as an example) total consumer benefit will be less. We also see from the figure that the required reservoir capacity is R*.

Insufficient reservoir capacity

If there is insufficient reservoir capacity, i.e. R < R*, there will be different prices in the two periods. That situation is shown in Figure 6.11.

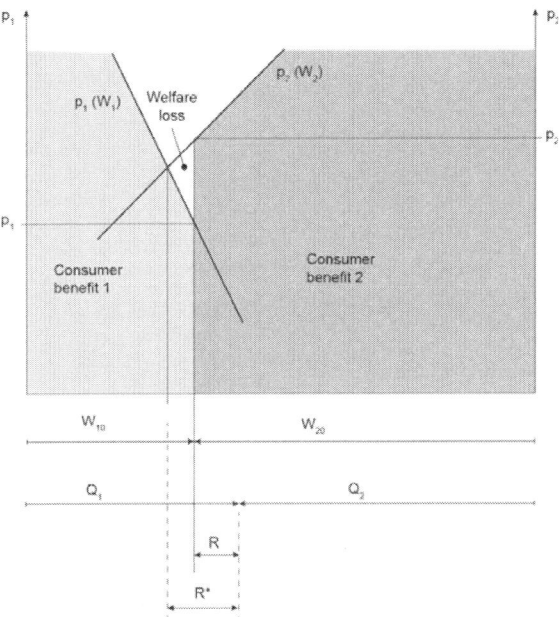

Figure 6.11. Two-period scheduling problem with a binding constraint on reservoir capacity.

There is an increase of consumer benefit in period 1 and a decrease in period 2 compared to the previous case, but there is a net loss (welfare loss or loss to society which in this case means net reduction in consumer benefit) due to limited reservoir capacity as shown in the figure.

Monopolistic solution

So far we have assumed that the generator's objective is social optimum. If the generating company is in a monopolistic position and its objective is profit, the situation is different. In this case the company's objective is to maximize its revenue:

$$\text{Max}\left[\pi_1 + \pi_2\right] = \text{Max}\left[p_1 W_1 + p_2 W_2\right] \quad (6.20)$$

The marginal revenue in the two periods are:

$$MR_1 = \frac{\partial \pi_1}{\partial W_1} = p_1 + W_1 \frac{\partial p_1}{\partial W_1} \qquad (6.21)$$

$$MR_2 = \frac{\partial \pi_2}{\partial W_2} = p_2 + W_2 \frac{\partial p_2}{\partial W_2} \qquad (6.22)$$

The maximum total revenue is achieved when marginal revenue is equal in both periods. That is shown in Figure 6.19.

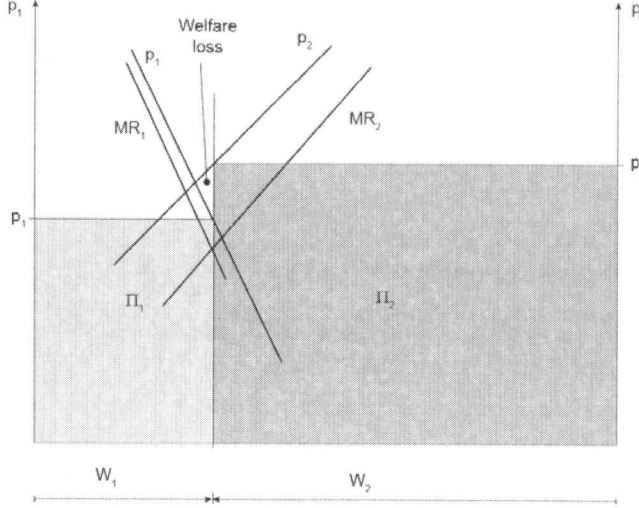

Figure 6.12 Monopolistic case.

We notice that there is some "overproduction" in the summer period so there is less water available for the winter. That leads to higher price and less production (but higher revenue) in the winter and a certain welfare loss compared with the optimal case.

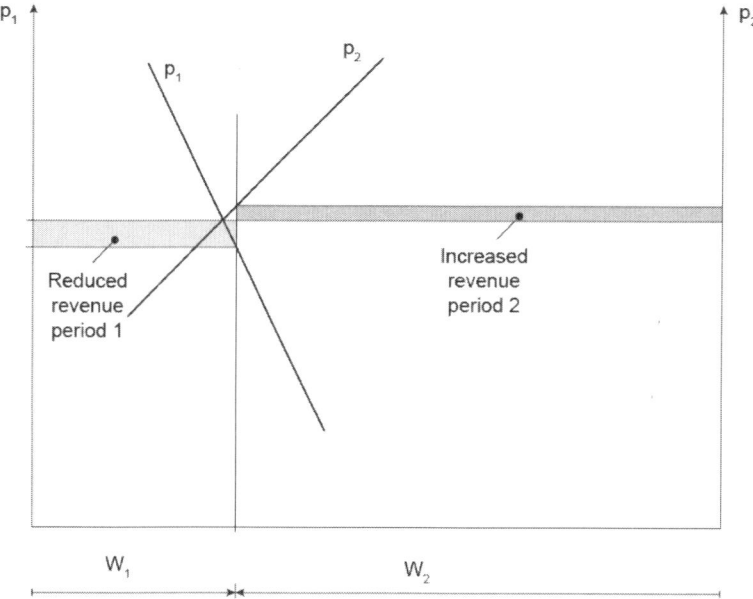

Figure 6.13 Monopolistic case. Change of revenue in the two periods.

It should be emphasized that this is the solution with one profit maximizing generator. With a number of competing generators, the solution will be close to the optimal solution, even with profit maximizing generating companies. If we assume the generating companies are price takers, they will act as if their own generation has no impact on the price, and that means:

$$\frac{\partial p_1}{\partial W_1} = 0 \Rightarrow MR_1 = p_1 \qquad (6.23)$$

$$\frac{\partial p_2}{\partial W_2} = 0 \Rightarrow MR_2 = p_2 \qquad (6.24)$$

That means we are back to the crossing point between p_1 and p_2.

Stochastic inflow

The inflow of water to a reservoir depends on weather conditions, notably precipitation and temperature. It is therefore a stochastic variable. The purpose of a reservoir is partly to serve as a buffer and absorb the stochastic variability of water inflow. Optimal use of the reservoir is still based on a requirement that the price should be equal in the two periods, but now we work under stochastic conditions.

In order to make optimal use the water under these stochastic conditions, we introduce *water values*.

A general comment to water values: The value of water is not tied to the cost of providing water or providing inflow to the reservoirs. The water is provided through precipitation which is free of charge.[5] But as long as the amount of water is limited, it will still have a value in terms of an *opportunity cost*. If we use it today, we will lose the opportunity of using it at a later stage. But in any case the water value is connected to the price in the power market.[6] The water value depends on the quantity of water available (large quantity gives low value) and the power demand in the market (high demand gives high value). Figure 4.2 gives an illustration of this general dependency.

If we look on our two-period model above and stick to the assumption that the inflow is a deterministic variable, the price can be different in different times of the year, but it will not depend on reservoir content. The reservoir will be filled up in the summer period and emptied in the winter. If the reservoir capacity is high enough, summer and winter prices (values) will be equal. If not, the winter prices are higher.

With a stochastic representation of the inflow (which of course is the correct representation), the water values depend on the time of the year and the content of the reservoir. If future inflow is high, the price we obtain in the future is low. If we start with a high reservoir level we can even risk overflow if future inflow is high. In case of overflow the value of water is zero. So generally the water value is low with a high reservoir level and vice versa. Under these stochastic conditions, the water value represents the <u>expected</u> value of a marginal amount of water if it is stored for later use. This expected water value is calculated based on knowledge about the probability for different inflow alternatives.

---

[5] There are of course investment costs connected to the building of dams, tunnels etc. But as long as we look on operational aspects, that must be regarded as *sunk costs* and are irrelevant. That type of costs are not affected operational decisions.

[6] If water is used for only one single purpose (electricity generation) which is the case in Norway the water value is only affected by the power market. In other countries water can be used for other purposes, for instance irrigation.

For our two-period model the water values can look as indicated in Figure 6.14. The water values are higher in the winter than in the summer period for a given reservoir level.

We start with the assumption that the water values are known in advance. In our model we are assuming that the decision concerning generation is taken at the beginning of each period and information about the inflow is available at the end of the period. This is of course unrealistic, but it explains how we in principle can make rational decisions confronted with an uncertain future.

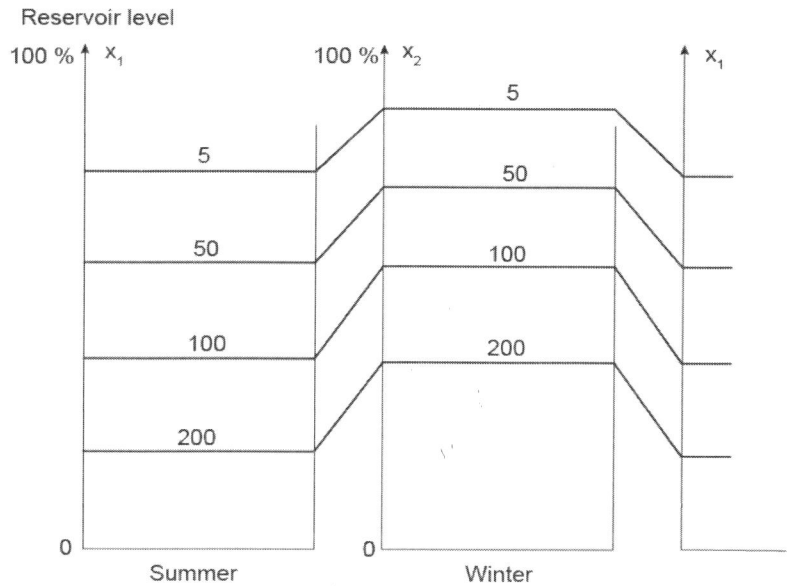

Figure 6.14. Water values represented by iso-price curves.

In Figure 6.15 we try to explain the link between the reservoir level with a corresponding water value and generation level in the two periods. The figure is divided in two parts: The upper part (Figure 6.15A) shows an illustration of the demand in the two periods similar to the previous illustrations. But it is no longer a bathtub diagram due to the stochastic inflow, there is no longer a fixed distance between the two vertical axes. The stochastic feature of the inflow is indicated by three different inflow scenarios for each period. The probability of each inflow scenario is assumed to be known. The lower part (Figure 6.15B) shows the reservoir levels and the corresponding water values, basically the same as Figure 6.14.

We start with a reservoir level x1 and a price p1 as indicated in B. This price is used to decide the generation level W1 as shown in A. Next, as it is shown in B, this generation reduces the reservoir level. At the same time the inflow fills up the reservoir. In our case we have three different inflow scenarios leading to three different reservoir levels, each with a known probability. We assume that the water value as a function of reservoir content is known and that makes it possible find the value for the three different scenarios. Combined with information about the respective probabilities, it is then possible to find the expected water value at the end of period 1 which means the start of period 2. Notice that we find the water value for each inflow scenario and based on that we find the expected water value. That is different from the water value at the expected reservoir level. We next repeat this procedure for period 2. We start with the price p2 from B and find the corresponding generation level in A. That leads to reduced reservoir level as shown in B at the same time as the reservoir is filled to three alternative levels depending on the inflow scenario. We can now find the expected reservoir level and the expected price at the end of period 2 with the same procedure as used in period 1. The result is supposed to match the reservoir level and the water value we started with in period 1.

In fact, the procedure described here is the procedure for calculation of water values. If there is a mismatch between the water value at the end of period 2 and the value we started with at the start of period 1, we have to make a correction and that is then a part of the iterative procedure we go through in our calculation of water values. If the iterative procedure has converged, there will be no mismatch.

If we are then using these water values in our regular decisions concerning generation, we will make sure that the market price we obtain by producing and selling a marginal amount of energy, will be compared to the expected value of storing that energy to a later period. If the market price is higher, we should produce, if it is lower we should hold back and store it for a later case. If our inflow has the same stochastic characteristics as the inflow scenarios used in the water value calculations, that will lead to an optimal use of water. Optimal in this context means maximum expected social surplus.

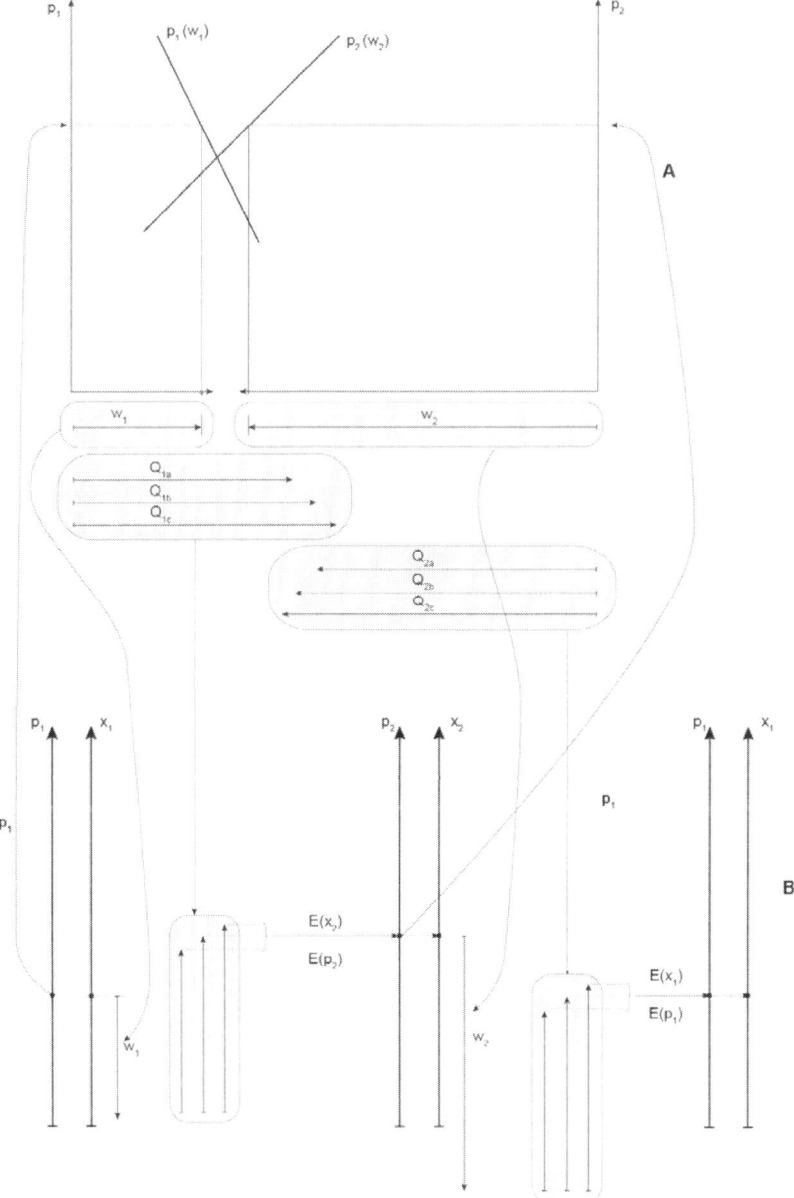

Figure 6.15. Illustration of a two period scheduling problem with stochastic inflow.

Practical water value calculation

The practical procedure for water value calculation is like the one just described, but the time periods are much shorter and the reservoir is divided into smaller fractions than indicated in Figure 6.15 We are normally using one week as time resolution and the reservoir is divided into 50 different levels. That means we have to find 52 x 50 different water values.

More information can be found in [5] and [6].

The hydro inflow to the reservoirs is variable over the year. The power demand is also variable. See Figure 6.16. This has to be taken into account in a practical procedure for water value calculation.

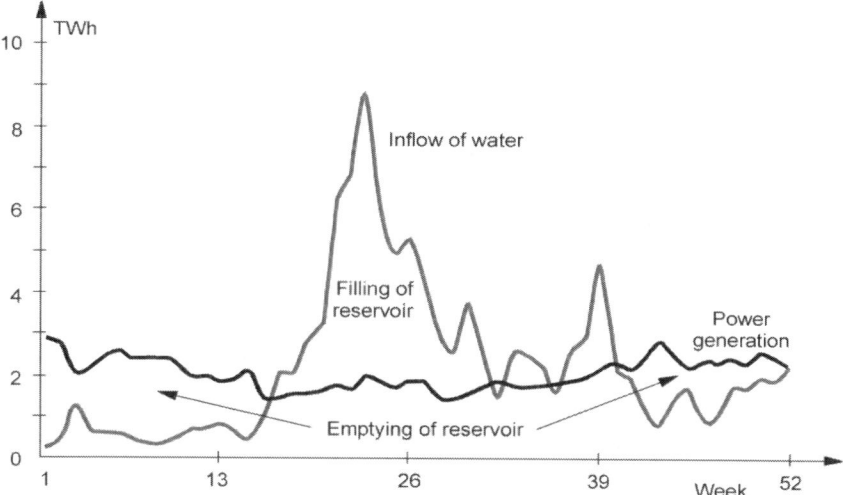

Figure 6.16. Inflow of water and power generation in the Norwegian system.

We have to calculate of a water value matrix with 50 x 52 elements. The procedure is indicated in Figure 6.17 below.

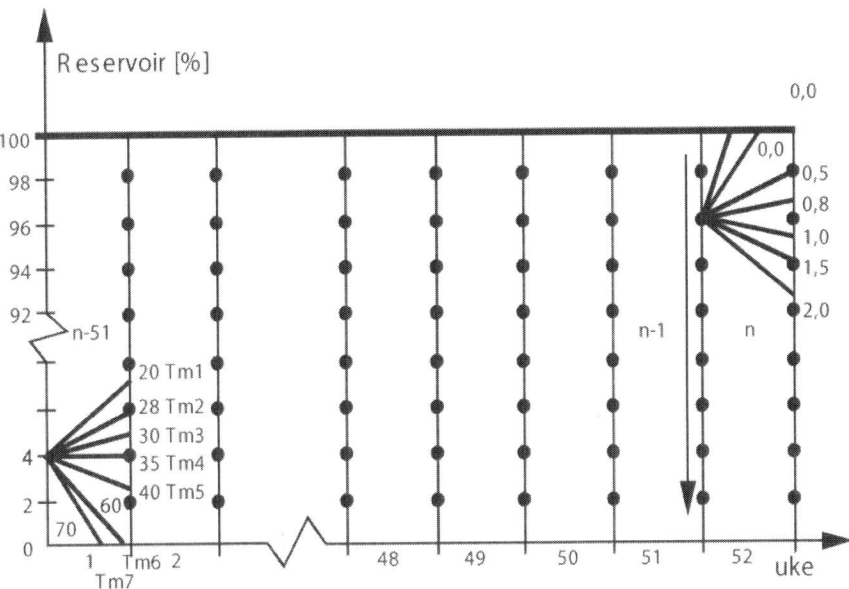

Figure 6.17. Procedure for calculation of water values over the year

The principle is the same as explained earlier. We start with a set of assumed water values, for instance for week 52, reservoir contents 96 % as seen in the figure. Given that value, we decide the generation level in principle as indicated in Figure 6.15. We then run through a set of inflow alternatives (7 is normally used) resulting in a set of alternative water values depending on the reservoir level we end up with. The expected water value can then be found based on probabilities tied to each inflow alternative. This is the new water value replacing the assumed one (or the one from earlier iterations). This procedure is repeated for every discrete level in the reservoir (50) and then we continue to week 51. The same procedure is then repeated until we reach week 1. But the water values at that point should equal the values we started with. If there is a difference, corrections must be made and we go through another iteration.

The results of the calculations can be as shown in Figure 6.18.

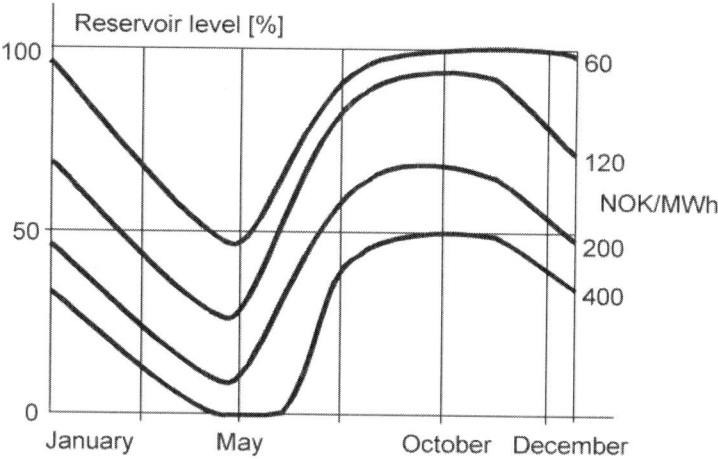

Figure 6.18. Incremental water values shown as equi-price curves.

6.4 Market modeling (EMPS)

6.4.1 Fundamental models

A generation scheduling model can be used as a market simulation tool. If the market is efficient, the optimality conditions are met and the result will be consistent with the output from the optimization model. These types of market simulation models, based on optimization and explicit description of the demand and supply side are called *fundamental models* in this context. We have other categories, for instance so-called *technical models*, which are also used for market simulation and forecasting, but we will not discuss such models here.

There are presently two models frequently used for the Nordic market: EFI's One area Power Scheduling model (EOPS) and EFI's Multi area Scheduling model (EMPS).

The EOPS is based on the principle for water calculation described above. The water value calculation is the dynamic optimization part of the model and the output - the water value matrix – represents decision rules for future decisions on production level. It is called the *strategy* part of the model.

After this strategy is established, the *simulation* part of the model can be run. This simulation is in principle quite simple. We run through a set of inflow scenarios. For each time step the decision on the production level is taken on the basis of water value (in line with the original basis for the water value calculation). With information on production level and the inflow, it is then possible to find the reservoir level at the next time step. In practice this simulation part can be complex because we often use a more complex representation of the hydro system than we use in the strategy part. The inflow scenarios are based on statistics from a long period. Today we normally use inflow data for 70 years (1930 – 2000).

Depending on the purpose of the simulation, we use two alternative approaches: *Parallel* or *series* simulation. We start with specified reservoir levels, for instance today's situation. Parallel simulation means that we run simulation from this starting point over and over again for different inflow cases. Series simulation means that we run one long simulation for all the years in series. After a few simulated years the model will then have "forgotten" the starting point. Parallel simulation, on the other hand, means that the starting point has more impact.

In most cases the EOPS model is too simple for a realistic simulation of the Nordic market. We have to use the multi-area model, EMPS. The EMPS model can be regarded as a set of parallel EOPS modules where there are possibilities for power exchange between areas.

For practical solution of the multi reservoir decision problem an approximate methodology is employed in the EMPS model. An SDP-related algorithm is used for solving each regional sub-problem, and an overlaying hierarchical logic is applied iteratively to treat the multi reservoir aspect. The process is illustrated in Figure 6.19

- Incremental water values are calculated for each region in a normal way by the use of backward SDP as described earlier.

- Simulation of total system behavior is next performed using the computed water values to determine energy generation in each subsystem, energy exchange between subsystems and transactions with neighboring regions.

- Feedback is then executed conditionally: If a stable and satisfactory solution is found, the process is terminated. If not, the result from the simulation is used to adjust regional premises, and return made to regional decision table computation.

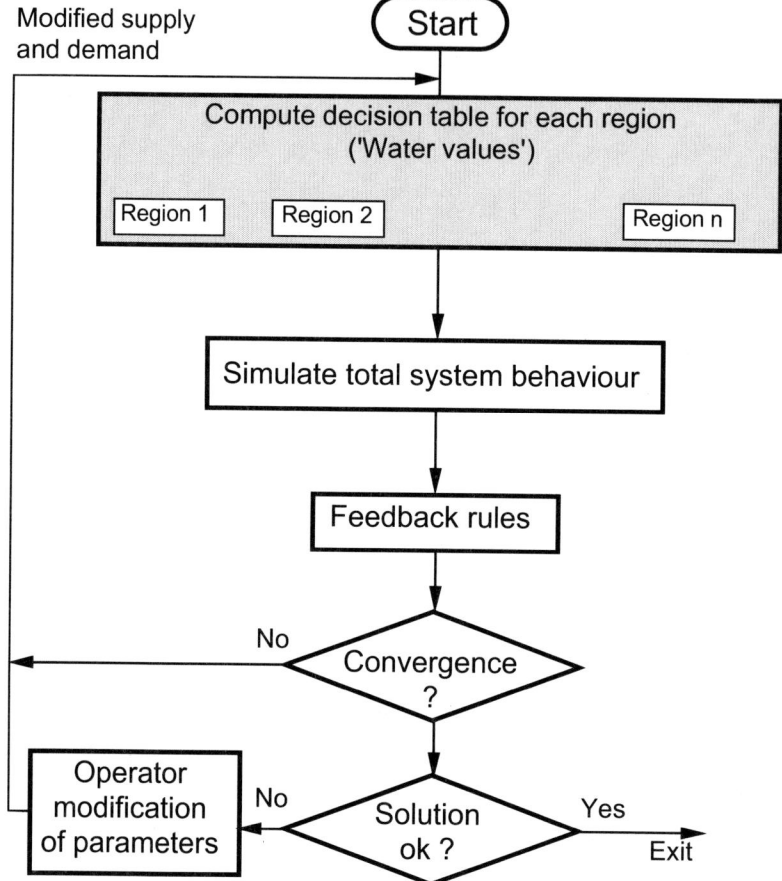

Figure 6.19 Main logic for solving the multi-reservoir problem.

6.4.2 The EMPS simulation part

In the simulation part of the EMPS model system performance is simulated for a chosen sequence of hydrological years. Based on the incremental cost tables calculated previously for each aggregate regional hydro system, weekly operational decisions on power generation (hydro or thermal) and consumption are made in a market clearing process. A detailed reservoir drawdown model gives the distribution of each subsystem's aggregated hydro generation among available plants each week. Historical inflow series covering a period of typically 70 years are normally the basis for simulation.

Market modeling (EMPS)

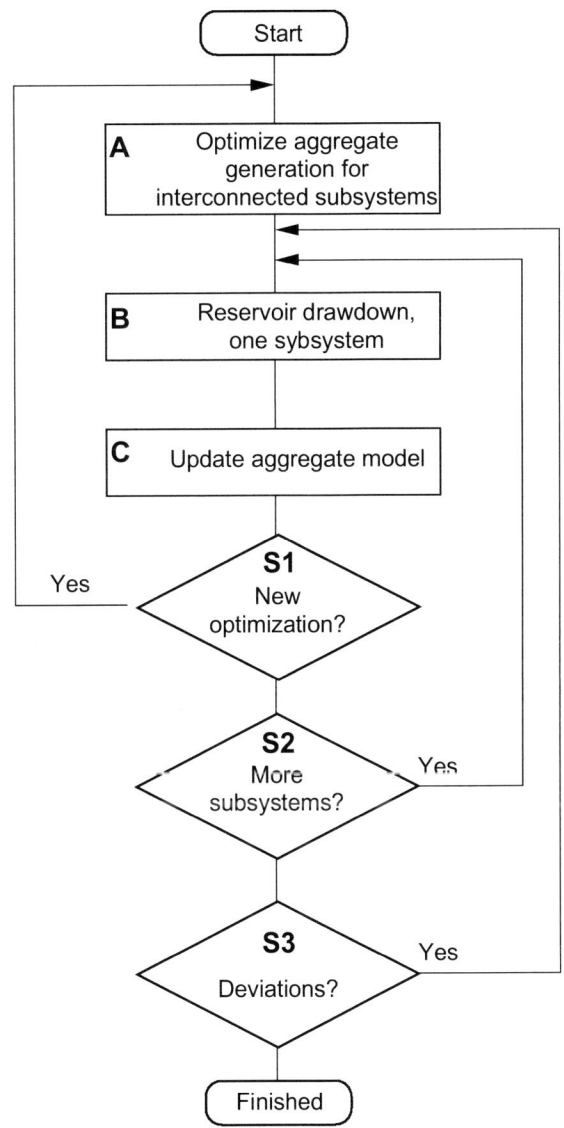

Figure 6.20 The weekly decision process in the EMPS model's simulation part.

Figure 6.20 illustrates the weekly operational decision process, summarized in the following points:

A Optimize regional (aggregate) generation
The optimum aggregate hydro generation is calculated, along with optimal thermal generation, consumption and exchange, given modeled incremental costs, capacities, losses and constraints. Hydropower is modeled as an aggregate equivalent reservoir, and an aggregate plant with a piecewise constant efficiency rate as well as a piecewise constant operational cost. This cost is added to the incremental value of stored energy.

B Reservoir drawdown
A detailed reservoir drawdown model gives the distribution of aggregate generation among available plants, and thus the distribution of stored energy among available reservoirs, according to a rule-based strategy.

C Update equivalent model
Upon exiting the reservoir drawdown model, the aggregate hydro model is always updated with the latest inflow, active generation constraints, pumping and plant efficiency. If this modifies the aggregate model in any way, optimum generation is recalculated by returning to *A*.

S1 New optimization?
The model returns to *A* for a new optimal decision whenever the equivalent hydro model has been modified, or if reservoir drawdown calculation is interrupted before completion.

S2 More subsystems?
Reservoir drawdown is performed for one subsystem at a time, and is repeated for all subsystems that have a local hydro system modeled in detail.

S3 Deviations in hydro generation?
This is basically a test on convergence.

Results that may be extracted from a system simulation with the EMPS model include:

- hydro system operation (reservoirs, flows, generation, pumping),
- thermal generation,
- power consumption, curtailment,
- exchange between subsystems,
- economic results,
- electricity-related emission figures,
- incremental benefit figures of increasing the capacity of various facilities (hydro, thermal, transmission system).

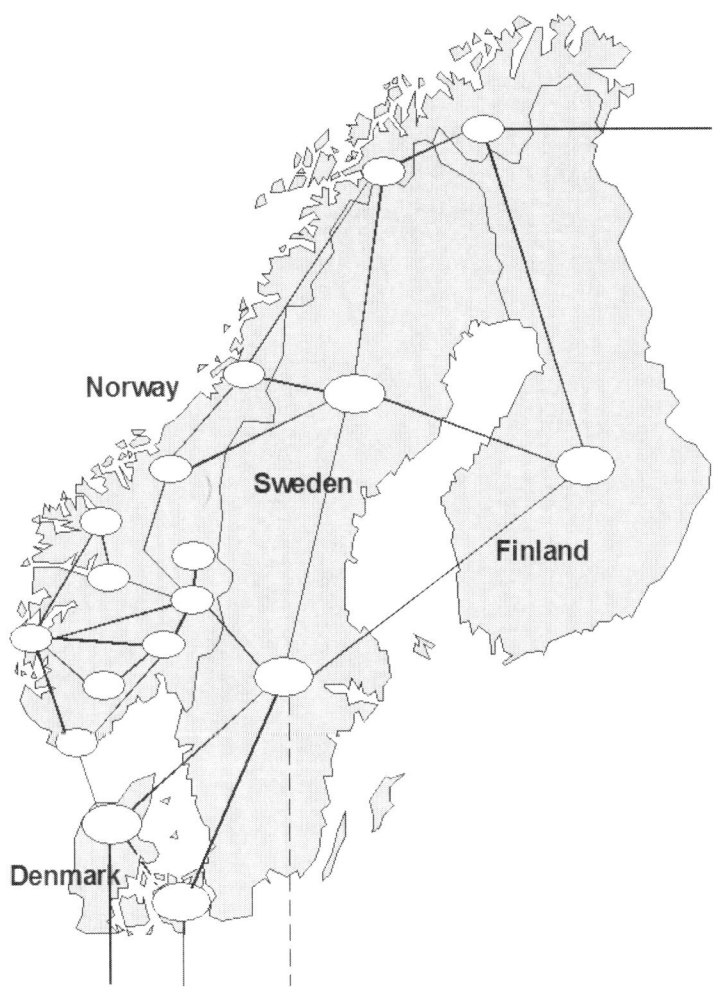

Figure 6.21 Structure of a frequently used version of the EMPS model

6.5 Generation planning in an open market

6.5.1 Introduction

So far this chapter has considered traditional generation planning, which means planning with the objective to maximize economic surplus. It has also described how this type of model can be used as a market simulation tool.

In this section generation planning in an open market is discussed. We assume that planning under the new circumstances is based on a decentralized system, where each generating company is planning its own production based on profit maximization, see Section 5.7. The basic difference between the two approaches is indicated by Figure 6.22. Under the former regime, consumption was regarded as given and we minimized the operation cost. The operation cost is represented by the area under the marginal cost curve. In an open market, the price is regarded as given (provided we have an efficient market) and the profit is maximized. Profit is indicated in the figure as the area between the marginal cost curve and the price line.

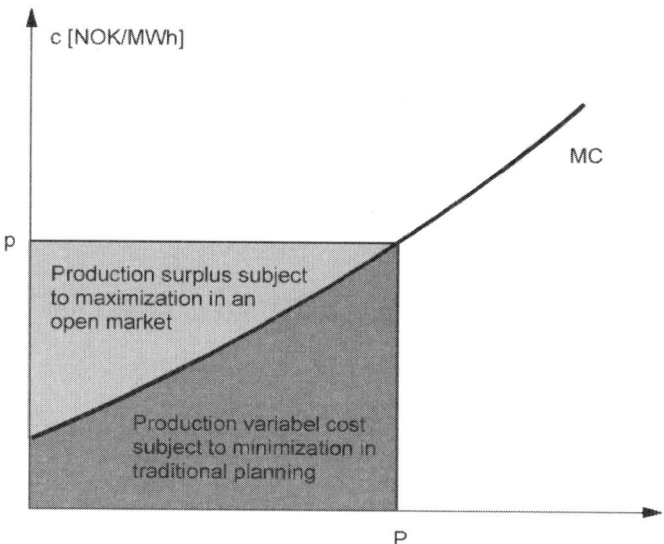

Figure 6.22 Difference between generation planning under the traditional and the deregulated regime

The figure indicates that the two approaches lead to the same result. But it should be underlined that this figure gives a simplified illustration.

6.5.2 Problem formulation in the traditional and deregulated environment

Traditional (Non-deregulated)

The following formulation represents the traditional (non-deregulated) description of the planning problem:

Given a forecast for (firm) electricity consumption: Establish a production plan (or strategy) that minimizes the expected cost of covering the consumption, all relevant constraints taken into account.

The cost (i.e. generation dependent cost) includes:

Generation fuel cost (which in Norway is zero in most cases)
+ costs for electricity purchasing (from the Pool)
- income from electricity sale (to the Pool)
- income from sales to interruptible consumption
+ curtailment costs.

This formulation is based on the former mandatory obligation of a utility to cover the firm power requirement within its concession area. If a consumer's demand for firm power is curtailed, the value of this is included in the curtailment cost. (The term: Value of Lost Load. VLL is also used in this connection).

Deregulated (or restructured) environment

In a deregulated environment the producers have in principle no obligation to serve any particular consumer. The only objective is to generate and sell electricity with maximum profits, which can be formulated in the following way.

Given a forecast of future market price (which in the long term is a stochastic variable): Establish a production plan (or strategy) that maximizes the expected profit over the planning period, all relevant constraints taken are into account.

The profit depends on:

Income from electricity sales (to the Pool).
Costs from electricity purchasing (from the Pool).
Generation fuel costs.

In this formulation of the planning problem it is assumed that the producers are *price takers* which implies a perfectly competitive market.

Deregulated versus traditional formulation

The planning formulations of the deregulated system differ from the traditional formulation in several ways:

We change from cost minimization to profit maximization. But the elements included in the profit function have opposite signs compared to the cost function, so what appears to be a difference, is in fact no real change.

Sales to interruptible consumption are not explicitly included in the formulation for a deregulated systems. Whether a consumer with an interruptible supply contract should be served or not, depends on the spot price and not on the running generation level. As long as supply to interruptible loads is assumed to be covered from the spot market, it is therefore not necessary to take this type of contracts into account in operation optimization. The same was in fact true in the former system, but perhaps not fully recognized.

Curtailment cost is not included in the new formulation, and this is in fact the only genuine difference between the two formulations. Curtailment cost is a real cost to society, but it is not included in the utilities accounts and does not affect the profit. Exclusion of the curtailment cost leads to a more risky operational strategy, i.e. a strategy implying an increased probability that a curtailment is needed. It is discussed to what extent this has really affected the Norwegian operational planning and the current security of supply. It has indeed been an increase of curtailment risk from 1990 until today, but this is mainly caused by the more narrow energy and capacity margin (which is also connected to the planning objective, in this case the expansion planning with the same basic mechanism). [12]

With the expected profit maximizing formulation the producer is regarded as a risk neutral agent. In real life that will not always be the case, but risk aversion (or a risk seeking attitude) is difficult to take into account explicitly in the problem formulation. However it can be modelled implicitly by introducing a penalty function. For more information see [9]

This problem formulation also implies that the producers are regarded as a *price takers*. That means that they do not take into account any influence their own production might have on the market price. This assumption seems reasonable for small producers, but may be questionable for larger ones. According to Nord Pool, 4 or 5 participants in the spot market have the capacity to influence spot prices. Given the participants' strong incentives to do so, it would be naive to think that those who are capable of it, do not at times consider their own influence on market prices when scheduling their generation or making bids. However we assume in the following that the producers are price takers.

6.5.3 Unit commitment in an open market

Unit commitment is a core element in traditional generation scheduling. That is based on a presumption that every unit is big enough to affect total cost and price.

In an open market we have to rely on the assumption that – in order to be efficient – not only every unit, but every generating company has to be small compared to the total market. In a perfect market, generation planning is based on the assumption that the price is given and we adjust the production volume in order to maximize profit. We will only start a unit if it is profitable, given the market price.

We go back to the examples from Section 6.3. Cost characteristics for the two generating units are shown in Figure 6.23. With the market price given, it is not profitable to start a unit until the price is above the AVC. For the two units shown, the price limit will be p_1 and p_2 and the units can be planned independent of each other. So in this static case the unit commitment problem is very simple.

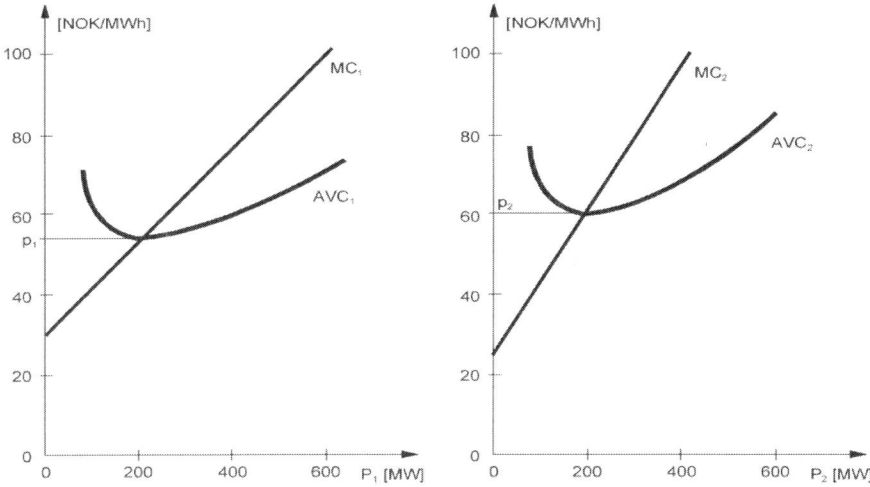

Figure 6.23 Cost characteristics for two generating units.

In a dynamic case, the unit commitment problem is not so trivial. Our planning is based on a price forecast, not on a given price at one point in time. Confronted with a price forecast, we can take start/stop costs into account. The accumulated profit during the period a unit is in operation must be higher than the start/stop cost. If the generating company relies on the price forecast, production from this unit can be bid into the market at marginal cost (or in principle at a price lower than the assumed market price during the period.) The bid will then be accepted and start/stop costs will be covered.

In the Nord Pool spot market the unit commitment problem is taken into account by the block bidding mechanism. A block bid includes a number of hours tied together. If a block bid for, as an example 4 hours, is submitted, the start/stop cost can be distributed over these 4 hours. So the bid price will be marginal cost plus distributed start/stop cost. If the bid is accepted, the full cost is covered. It is not a perfect mechanism because the choice of blocks is limited. But it is a support for generating companies, especially as long as they cannot rely entirely on their price forecasts.

Start/stop costs as well as restrictions can be taken into account in the scheduling as long as no price impact from individual units is assumed. We avoid the complexity of traditional unit commitment where all combinations of on/off states have to be included. That makes unit commitment much simpler under market conditions than in the traditional setting.

The unit commitment problem is to some extent ruled out by the free market conditions. In any case it is simplified and it can be solved independently for each unit. It can be argued that no market is perfect. Generation from one additional unit will have some impact on the market price. So what we are doing represents an approximation. Traditional generation planning is based on an ambition to reach the exact optimum. In an open market we must always accept an approximately perfect condition and we are content with an approximately optimal solution to the unit commitment problem.[7]

6.5.4 Hydro generation

Planning of hydro generation in an open market is based on the same criterion: maximization of expected profit. The objective is:

$$J(x,N) = E\left\{ Max \sum_{i=1}^{N} P_i \cdot p_i + S(x,N) \right\} \quad (6.25)$$

where generation level multiplied by price, $P_i \cdot p_i$ is the addition to the objective function in time period i. Notice the difference between this formulation and equation (6.19).

The profit is affected by two stochastic variables: market price and inflow of water. In Section 6.3 we went through a procedure for operation optimization with stochastic inflow on the input side and a price-quantity relationship on the demand side. We will here explain how this procedure is affected by the price taker assumption.

First we explain how the planning procedure is simplified if we assume that the price is given and constant whereas the inflow is stochastic. Next we go through an optimization procedure with stochastic price and given quantity. In the next section planning based on both stochastic prices and stochastic inflow is described. That description is tied to planning procedures and tools presently used in the Nordic market.

---

[7] We assume here decentralized scheduling. With centralized scheduling unit commitment can be included in line with traditional planning. See Section 5.7

Given constant price, stochastic inflow

If there is no price variation, the price is not affected by the producer's own generation and there is only a restriction on the generation quantity (generation capacity), the problem is only to avoid water spillage. The solution to this is to keep the reservoir level as low as possible all the time, which means maximum generation as long as there is water available. (We disregard how the reservoir level impact the head)

Given quantity, stochastic price

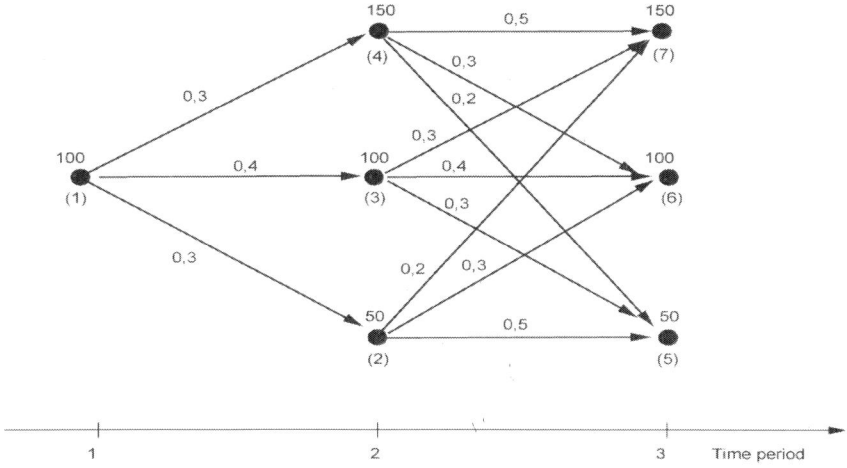

Event tree for possible future market prices. Each node represents a price and the arrows represent transitions with probabilities. The number in the brackets is used to number each node in the tree.

Figure 6.24 Stochastic price model.

We use a simple numerical example. It is based on the stochastic price model shown in Figure 6.24. We have three time periods. As indicated on the figure, we define a set of nodes marked with brackets (1), (2), ...(9), each tied to a price and a time period. The probabilities for transition from one node to another are indicated by the numbers on the arrows. The price in period 1 is 100 NOK/MWh. The price in period 2 is either 150, 100 or 50 with probabilities 0.3, 0.4 and 0.3 respectively.

For this simple case, we can describe the uncertainty with a limited set of scenarios for future price development, see Table 6.2

Table 6.2 Scenarios for price development. All prices i NOK/MWh

Scenario	Price in period1	Price in period 2	Price in period 3	Probability	Selling Price	Expected Income Mill. NOK
1 (Node (1)(2)(5))	100	50	50	0.3 ·0.5 = 0.15	50	0.75
2 (Node (1)(2)(6))	100	50	100	0.3 ·0.3 = 0.09	100	0.90
3 (Node (1)(2)(7))	100	50	150	0.3 ·0.2 = 0.06	150	0.90
4 (Node (1)(3)(5))	100	100	50	0.4 ·0.3 = 0.12	100	1.20
5 (Node (1)(3)(6))	100	100	100	0.4 ·0.4 = 0.16	100	1.60
6 (Node (1)(3)(7))	100	100	150	0.4 ·0.2 = 0.12	100	1.20
7 (Node (1)(4)(5))	100	150	50	0.3 ·0.2 = 0.06	150	0.90
8 (Node (1)(4)(6))	100	150	100	0.3 ·0.3 = 0.09	150	1.35
9 (Node (1)(4)(7))	100	150	150	0.3 ·0.5 = 0.15	150	2.25
Expected price	100	100	100	$\Sigma = 1.00$		$\Sigma = 11.05$

We notice that expected price in all periods is 100 NOK/MWh.

We assume than that we have a hydro plant with a reservoir of 100 GWh. This reservoir has to be emptied at the end of the three periods. There is sufficient generation capacity to produce 100 GWh within one period. The problem is scheduling.

If the water is not used in the first or the second time period it must be used in the third time period independent of the price.

In order to avoid complexity, discounting of future cash flows is not included. It is assumed that the owner of the plant operates as a price taker in a market. We also assume that the owner has a profit maximizing objective and is risk neutral. There are no constraints on transmission or other constraints. What is the value of the water and when (which time period) should the water be used?

It is assumed that the price in each time step is known before a decision is taken.

As we see from Table 6.2 the expected future spot prices seen from the start, is 100 NOK/MWh for both the second and the third time periods.

This problem can be solved by investigation of each scenario as shown in Table 6.2. We have 9 scenarios and the probability for each is shown in the table. If the price in period 2 is 50, it is profitable to store the water and wait for better prices. The selling price (in period 3) can then be 50, 100 or 150 NOK/MWh. For the other scenarios we will sell in period 2 and receive a selling price of 100 or 150 NOK/MWh. We find that the expected income is NOK 11.05 millions. So the water value seen from period 1, which means the value provided optimal dispatch, is 110.5 NOK/MWh. That means it is not profitable to sell out in period 1.

Generation planning in an open market

This water value by investigation of each scenario is only possible for simple cases. In practical planning a procedure based on stochastic dynamic programming (SDP) is often used. We will demonstrate that approach on this same case.

First the optimal decisions for period 2 are found[8]. We start in node number (2). The water value is found by calculating expected future income if the water is stored till time period 3. As we see from Table 6.3 the water value is 85. The current market price is 50, so it is unprofitable to produce and sell. We then proceed to node (3). In this case the water value is 100 and the market value is also 100. In this case we are indifferent between selling and storing. Selling is most attractive for a slightly risk averse decision maker. Node (4) gives a water value of 115 whereas the market price is 150. It is profitable to produce and sell.

Table 6.3 Table for calculation of expected future value of water in time period 2

Node	Water Value (expected future value of water)	Market price	Production/sales
(4)	0.5 → 150 0.3 → 100 0.2 → 50 WV = 0.5 · 150 + 0.3 · 100 + 0.2 · 50 = 115	150	100
(3)	0.3 → 150 0.4 → 100 0.3 → 50 WV = 0.3 · 150 + 0.4 · 100 + 0.3 · 50 = 100	100	100*
(2)	0.2 → 150 0.3 → 100 0.5 → 50 WV = 0.2 · 150 + 0.3 · 100 + 0.5 · 50 = 85	50	0

---

[8] We call this backward SDP because we start at the last time step and move backwards.

Chapter 6: Power generation

A risk neutral producer will be indifferent to producing or not in this case. The water value is equal to the market price. But a slightly risk averse producer will produce and sell.

Table 6.3 shows how water values in time period two for all nodes in that time period ((2), (3) and (4)), are calculate assuming that the contract is not already used.

The corresponding calculation of water value for the first time step is shown in Table 6.4. We are still calculating expected future values based on to the assumption that optimal decisions in the subsequent time steps are taken.

Table 6.4 Calculation of expected future value of water in time period 1

Node	Water Value (expected future value of water)	Market price	Production/sales
(1)	0.3 → 150 0.4 → 100 0.3 → 85 WV = 0.3 · 150 + 0.4 · 100 + 0.3 · 85 = 110.5	100	0

Table 6.4 shows that the optimal decision is not to produce and sell in the first time period. The expected value of the water is 110.5 NOK/MWh while the market price is 100 NOK/MWh (the same as the expected price in the second and third time periods).

The difference between 110.5 and 100 NOK/MWh represents the value of having flexibility in the use of water. Instead of selling in period 1, it is possible to increase expected profit by a "wait-and-see" strategy. There is a certain chance that the price will rise. In that case we produce and sell with a good profit. And if the opposite happens - the price falls - there is still a chance will rise again if we wait until the third period.[9]

Notice the similarity between this water value calculation and calculations based on different inflow scenarios. Figure 6.15 shows how water values represent the expected future value of water given a set of inflow alternatives with known probabilities. Table 6.3 and 6.4 shows how water values are calculated based on at set of future market prices with known probabilities.

---

[9] This is an example of practical use of real options theory. The difference between 11.05 and NOK 10.0 million represents the <u>option</u> value of having flexibility in the use the water.

6.6 Generation planning in practice

6.6.1 Overview

Figure 6.25 gives an overview of the activities of different entities and links between them.

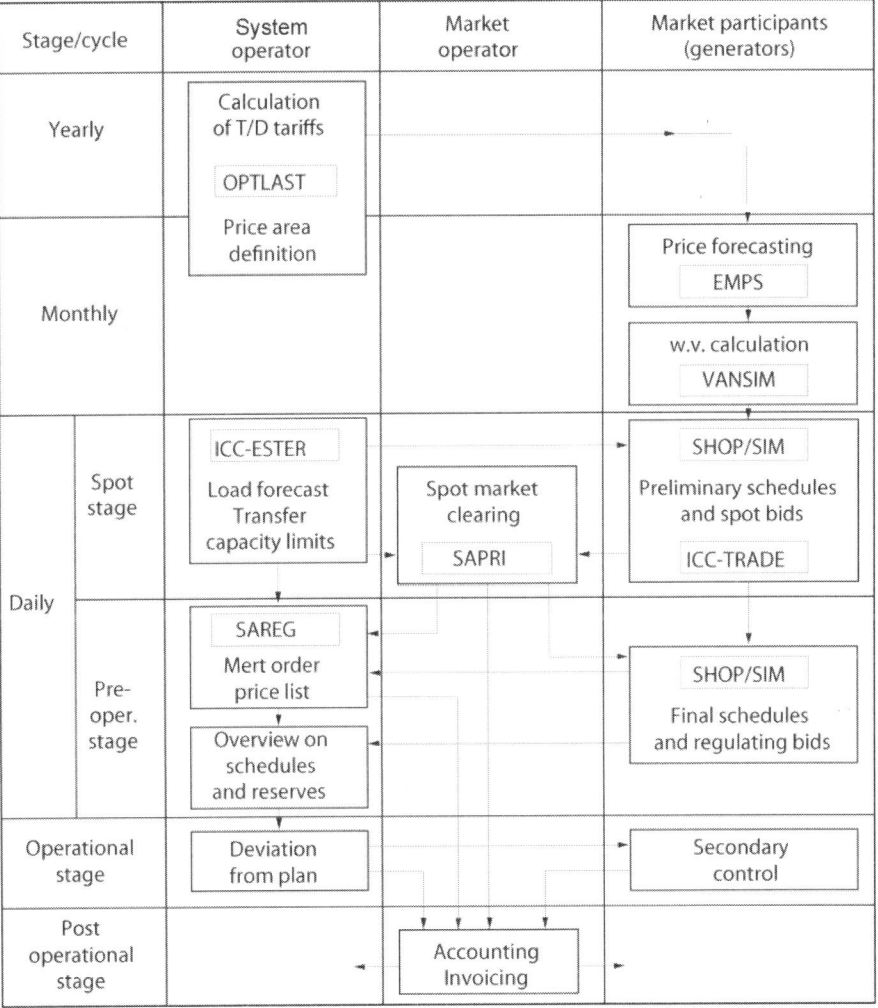

Figure 6.25 Activities of different entities and links between them.

Market participants are buyers and sellers in the market place. These can be generating companies, distribution utilities or end users. But here we focus on generators.

The market operator or *exchange* (Nord Pool) is responsible for the market clearing process in the spot market (24h market) and the Futures market. Accounting and invoicing is also the responsibility of the market operator.

The system operator, which in Norway also is the operator of the central grid, is Statnett.

Operation planning[10] is performed in a dynamic interplay between the three main parties in the system: 1) The market participants on the supply and demand sides, 2) the System Operator and 3) the market operator as indicated in Figure 6.25.

The first action taken, that affects the operational decisions, is *price area* definitions. This is done by the System Operator. Certain price areas will maintain for several years, others will be adjusted more frequently.

The price areas are defined in such a way that lines that are expected to be overloaded, are connecting different areas. The System Operator is also responsible for deciding the transfer limit on all transmission lines in the main grid. Spot market prices are calculated for each price area, and the differences in area prices represent the congestion fees as explained in more detail in Chapter 7.

The System Operator is also responsible calculation of transmission tariffs for the central grid. This is normally done once a year for the operational independent components of the tariffs. In Norway the operation dependent components (energy terms) of the central grid tariffs are presently recalculated every week. (See Chapter 7 for more details).

A crucial activity in the scheduling process is price forecasting. It is left to each participant (generator) to make a forecast for the prices at the nodes where the producer's units are connected. This nodal spot price depends on the general market balance, i.e. the *system spot price* and transmission charges including possible congestion fees.

Scheduling is broken down in long term (3 - 5 years ahead), medium term (1 - 2 years) and short term (1 week) scheduling, as will be described in more detail below, (see also [5]). These details are shown in Figure 6.26

As explained earlier, generation scheduling is based on the following problem formulation: Given a forecast of future market price (spot price): establish a generation schedule (or strategy) that maximizes expected profit over the planning period.

---

[10] In this overview we are not including maintenance planning. The System Operator plays a major role in coordinating maintenance, but this is not included in this description.

As long as the objective is *expected* profit, the scheduling can be done as if all generation is sold on the Spot Market. Any marginal quantity will in any case go to that market. If we extend the formulation to include (financial) risk, it is necessary to take the contract portfolio into account.

The spot price depends on several factors, but the most important is the amount of stored water in the total system. Since there is a slow variation reservoir contents, the spot price has a strong sequential correlation.

6.6.2 Long-term scheduling

Since we are dealing with a single company here, we follow the traditional way of aggregating the system into a one-reservoir model for which an operating strategy can be found by stochastic dynamic programming. In long-term scheduling the inflow must be dealt with as a stochastic variable. What is characteristic for the present approach, is that the price is also modeled as a stochastic variable.

Forecasting future spot market price

A forecast of future market prices is needed in order to obtain an optimal operational strategy. This forecast must include not only an estimate of the expected price, but also a description of the distribution.

In principle there are three different approaches to the forecasting problem:

1. One can use prices from the futures market. One of the drawbacks of this approach is that this price contains little information about correlations and price uncertainty, both being important factors for planning purposes.

2. One can use observed prices in the past and make forecasts based on trends and patterns in these historical observations. However, a reliable estimation of the stochastic parameters needed, requires prices from a long historical period where the total supply and demand is close to the expectations of the planning period (2-3 years ahead). As long as the supply and demand are growing, such observations are not available.

3. One can use a simulation model that describes the price formation.

In order to avoid the drawbacks of the two first methods, many generating companies often prefer alternative 3 and use EMPS-model as indicated in Figure 6.27. Parallel simulations are used in this context. It is also possible to use a combination based on alternative 3 with adjustments/ corrections from the other two.

Chapter 6: Power generation

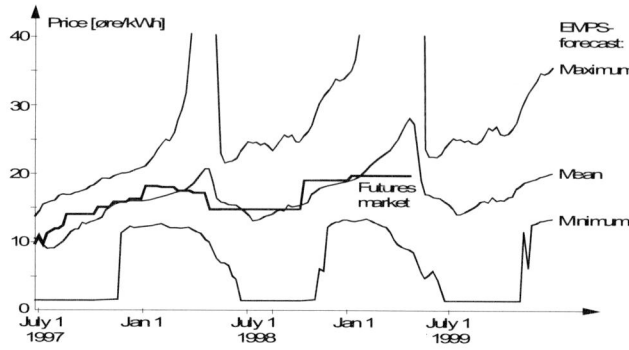

Figure 6.27 EMPS-forecast compared with price on the Futures market.

The experience with these model-based price forecasts is reasonably good; the model predicts the probable range of future prices fairly accurately, but only provided it is supplied with sufficiently accurate predictions of power demand, initial reservoir storage (including at times water stored as snow) and the behavior of foreign trading partners.

Stochastic model for the market price

The analysis of the price scenarios as well as observations of real spot price variability shows that each week's market price strongly depends on the price of the foregoing week. In order to use an optimization algorithm based on SDP, we need to establish a stochastic model of the price variations where this dependency is taken into account. The model we establish is in principle the same as the one shown in Event tree for possible future market prices. Each node represents a price and the arrows represent transitions with probabilities. The number in the brackets is used to number each node in the tree.

Event tree for possible future market prices. Each node represents a price and the arrows represent transitions with probabilities. The number in the brackets is used to number each node in the tree.

. The procedure for establishing this model on the basis of price scenarios from the EMPS model is described in the Appendix to this chapter.

Stochastic inflow model

Inflow to the aggregate hydro model is calculated by using a detailed scheduling model to simulate system operation for the time span where inflow statistics are available. We use the one-reservoir model, see Figure 6.5. Water which for some

Generation planning in practice

reason cannot be stored is included in the uncontrollable inflow to the aggregate plant. That includes:

- Inflow to run-of-rivers plants.
- Generation resulting from minimum flow constraints.
- Necessary generation to avoid immediate spillage.

All other inflow, minus spilled water is regarded as storable inflow.

Inflow one week is to some extent linked to inflow the previous week, but the sequential correlation is not nearly as strong as in the case with market price. For water value calculations we use an autoregressive (AR) model.

Each week is treated independently. A weighting procedure is applied to calculate the weights assigned to each of the 7 weekly scenarios in order to assure that accumulated inflow over the year have the correct expected value and standard deviation in the water value calculations.

Local inflow is treated as being independent from one week to the next (which is incorrect, but partly compensated for by the weighting procedure) and independent of market price.

6.6.3 The medium-term scheduling model

The planning period for the medium-term model starts at the end of the long-term planning period. The reservoir content at the beginning of the planning period is assumed to be known. The length of the medium-term planning period may be up to 18 months. The reservoir content is given by the simulated results from the long-term planning model. Failure to meet the targeted reservoir at the end of the medium-term planning period is penalized.

The medium-term model is a deterministic optimization model based on network flow programming. The uncertainty in inflow and market price are accounted for by optimizing for a number of scenarios, the same scenarios as used in the long-term model.

The main results from the model are the incremental water values for each reservoir in the system and the sales revenues for the planning period.

The coupling method between the medium-term and the short-term planning models, requires that the calculation for all the scenarios are made for a number of combinations of initial reservoir contents. These initial volumes are chosen in such a way that they reflect all possible end storages for short-term scheduling.

6.6.4 The short-term scheduling model

The purpose of short-term scheduling is in general to optimize the balancing of supply and demand in the near future while adapting to the long-term strategies for system operation. The short-term model gets its boundary conditions from a medium-term model. The time resolution of the medium-term model is one week and the short-term model must connect with the time intervals where the boundary conditions are available.

Reservoir trajectories for different inflow scenarios and incremental cost descriptions for individual reservoir volumes are typical results from medium-term scheduling. The inflow for the coming week will always deviate from the assumptions made in medium-term scheduling which usually does not have the same details as the short-term model. It is therefore not suitable to use those reservoir trajectories as boundary conditions when this may lack the flexibility necessary for handling variations in local inflow and deviations between short- and medium-term system models.

The short-term model must take into account a complex configuration of cascaded reservoir systems with different storage capacities and operational constraints on discharge levels, rate of change between successive time intervals and reservoir levels at user specified time intervals while adapting to the boundary conditions given from medium-term scheduling. The time resolution is usually one hour. Demand can be given as a market description, a firm load obligation or any combination depending on which sub-interval of the study period and application the program is used for. The case to be solved is therefore both hydraulically coupled and time constrained and thus represents a large scale optimization problem.

The program used to solve this problem [11] is based on successive linear programming. It uses a detailed model of the hydro system including basic modules such as reservoirs, power plants and discharge gates. The reservoirs are the basic connecting elements in the program formulation, and a plant connecting two reservoirs can consist of a tunnel branching into several pipes where each pipe may have several units. A non-linear representation of the efficiency curve for the individual hydro units is used. The plant efficiency curves as a function of upstream and downstream reservoir elevation are calculated based on a best efficiency strategy of loading the units while accounting for head loss tables for the tunnel, head loss coefficients of pipes and tailrace effects.

The short term model is used for different purposes. It is used as decision support for short-term bids. In addition it is used to calculate the final schedules after the power exchange obligations are set in the market clearing process.

6.6.5 Details about the bidding and price clearing process

Figure 6.28 shows the daily bidding and price clearing process in detail.

---

[11] The program referred to is developed by SINTEF Energy Research and is called SHOP (Short-term Hydro Optimization).

Generation planning in practice

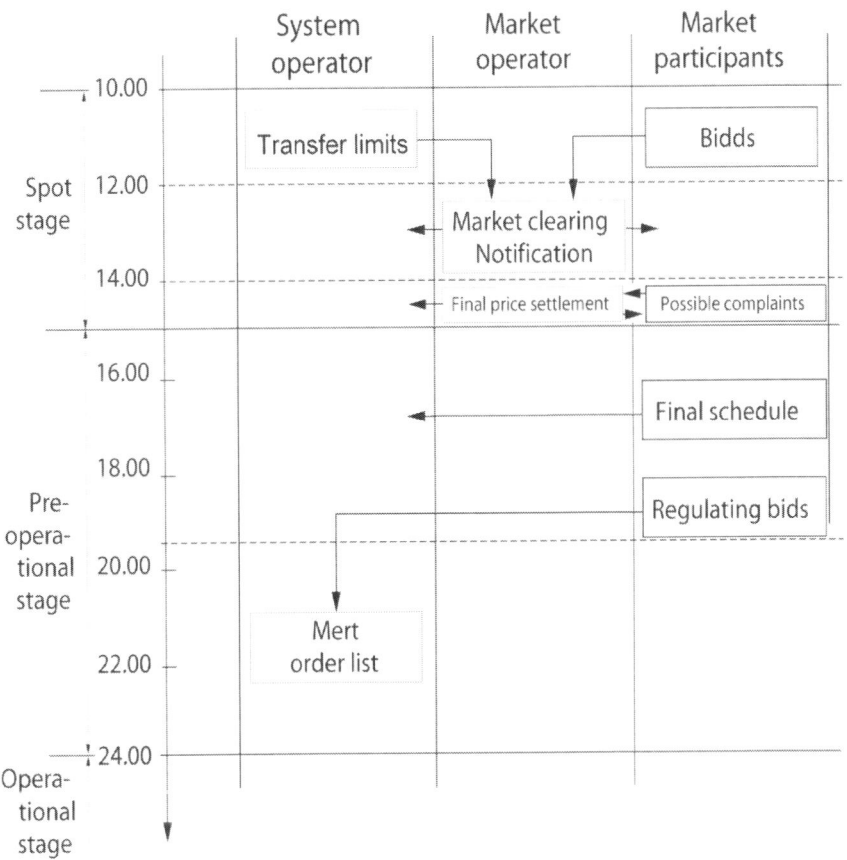

Figure 6.28 Bidding and price clearing process.

Spot market bidding

The bidding process goes on until 12.00 am. A standard form is used. (see Figure 6.29). The bidding can be done by fax or, most preferably, by electronic data communication (EDK). A participant can as indicated in Figure 6.29, submit both sales and purchasing bids at the same time. In this case the participant is willing to buy if the price is under 170 NOK/MWh and sell if the price is over 170 NOK/MWh.

The bidding is referred to each price area. This means that at this stage there is no information available to the market- or system operator concerning the individual generating units.

Bidder: Storfossen AB
Week no.: 40/95
Price area: C

Hour From-To	Price									
	0	100	101	150	151	170	171	200	201	900
1-6	50	50	20	10	0	0	-10	-20	-30	-30
7	80		80		30	0	-20	-30	-30	-20
8	80			80	50	0	-20	-20	-30	-30
9	80			80	70	0	-20	-20	-30	-30
10	80				80	0	-10	-10	-30	-30
11-16	100	100	100	100	80	0	-10	-10	-30	-30
17	80	70	70	60	50	0	-20	-30	-30	-30
18	80	70	69	20	19	0	-20	-20	-30	-30
19	80	70	60	40	0	0	-20	-20	-30	-30
20	80	70	40	30	0	0	-20	-20	-30	-30
21	70	70	30	20	0	0	-20	-20	-30	-30
22-14	50	50	20	10	0	0	-10	-20	-30	-30

Shading indicates 80 MW

Figure 6.29 Bidding from 24h market (NOK/MWh and MW)

Price settlement

The price settlement is done simply by adding all the bids on the demand side and the same on the supply side and then finding the point where the supply and demand curves[12] intersect. When this is done for the whole market, with no regard to the price areas, the computed price is the price we would obtain if there was no congestion. This is called the *system price*.

Next the power balance in each price area is considered. The bids contain information about the price area and Nord Pool can balance sales and purchase bids and find deficit or surplus (provided the system price is prevailing) for each price area. If that gives an indication that there will be an overload on the line(s) between two price areas, the area prices are adjusted in order to keep the transfer within the capacity limits. The congestion problem is discussed in more detail in Chapter 7.

---

[12] This supply/demand curves are of course entirely different from the "traditional" supply/demand curve we would have if all the generators had been on one side and all the consumers on the other. In this case there is a mix of generation and consumption on both sides.

The preliminary price settlement is finished and the participants are notified by 14.00. Possible complaints can be sent to the market operator within the next halt hour, and before 15.00 the final spot prices are settled and published. This spot price is available on Internet a few minutes later.

Final scheduling

After the spot prices are settled, the final production schedule can be worked out. Normally the bidding process will include a preliminary scheduling. As indicated in Figure 6.30, bids and prices refer to the prevailing price areas in what we call *the market phase*. In the *control phases* - the pre operational and operational phase – there is a nodal resolution which means that schedules and regulating bids are referred to their individual nodes.

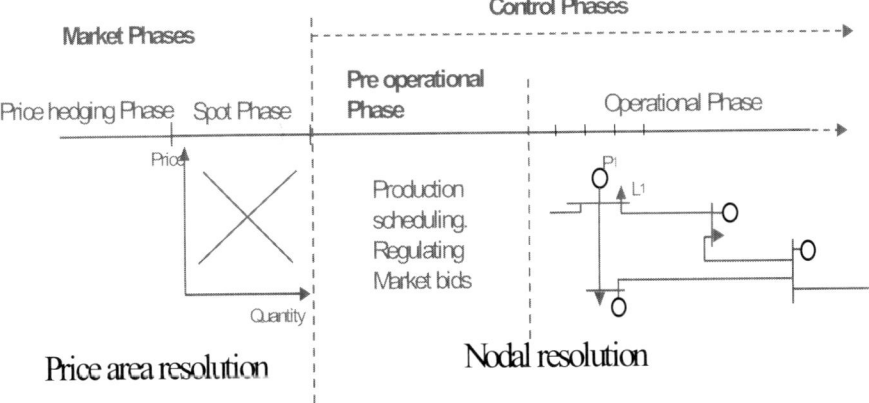

Figure 6.30 Market and control phases

6.6.6 The Balancing Market

After the spot market is closed, the Balancing Market (or Regulating Market) is opened for bidding. As indicated in Figure 6.30 the spot market is tied to price areas. Bidders in the Balancing Market must specify the node the objects are tied to.

The Balancing Market serves as a tool for System Operators to balance power generation to load at any time during real-time operations and provides a price for participants' power imbalances. The BM is thus a collection of regulating objects to compensate for any imbalance between production and demand during the operating phase. Bids in the balancing market are submitted to the TSO after the Elspot market has closed. Bids may be posted or changed close to the operational time, in accordance with agreed rules. Demand-side bidding for both increased or reduced consumption, as well as supply-side generation, are posted to the market. During recent years the Nordic TSOs have co-ordinated the balancing markets. Presently there is a common Nordic system with available regulating bids, and it is applied, largely, identical pricing principles for active regulations. These changes in essence cause the TSOs to together focus on real-time balancing of the overall Nordic grid rather than on separate balancing of the national sub-systems. Among other things this could lead to improved efficiency because of potentially decreased overall regulating power volumes and more competition between players providing regulating bids.

Balancing market bids are for upward regulation (increased generation or reduced consumption) and downward regulation (decreased generation or increased consumption). Both demand-side and supply-side bids are posted, stating prices and volumes.

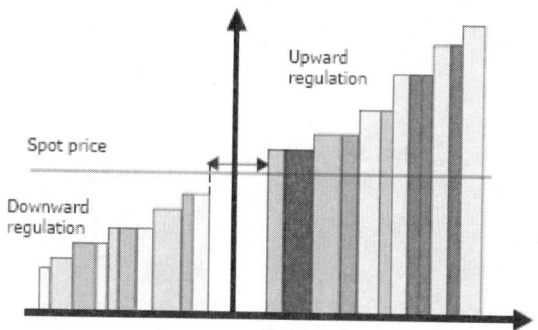

Figure 6.31 Bids in priority order for each hour

There are bids for increased generation or reduced consumption, the participant requires a price to increase generation or reduce consumption for a specified volume. Lowest bid price is the Elspot price. But also for decreased generation or increased consumption bids, the participant offers a price to pay to reduce generation or to increase consumption. Highest bid price is the Elspot price.

After operation is finished, all metered data are collected and the imbalances for all participants are calculated. The individual participant is charged or credited for his power imbalances, based on the prices in the Balancing Market.

Optimal mix of generating units.

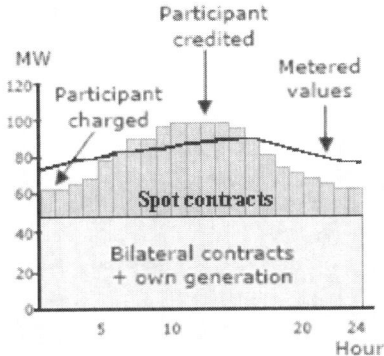

Figure 6.32 Graph explaining pricing of participant's imbalances

6.7 Optimal mix of generating units

We will limit the discussion of investments and investment planning in this book to the description of a procedure for determination of optimal mix of generating capacity. It is a so-called screening test based on equilibrium conditions.

6.7.1 Optimal mix based on cost minimization

The procedure is based on the following assumptions:

Given a yearly electricity consumption described by a load duration curve. The load is assumed to follow the same variation pattern from year to year. It is static in the sense that there is no change – increase or decrease – from one year to the other. The available generation technologies are characterized by two cost parameters: a fixed yearly cost [NOK/kW year] covering capital cost and fixed operation cost and an operation cost [NOK/kWh] covering fuel costs mainly. In this simple analysis we disregard hydropower.

The principle for the technique is shown in Figure 6.33. The load duration curve is given. We assume it is not affected by price or other factors.

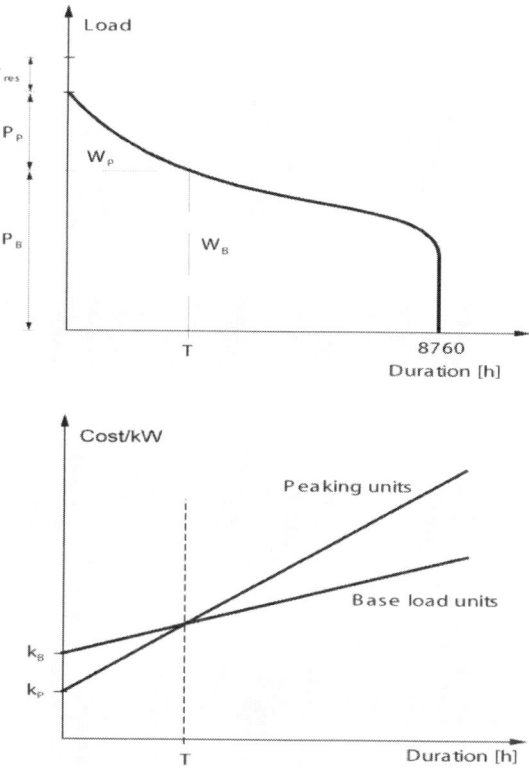

Figure 6.33 Optimal mix of generating capacity.

In order to make the example simple, we consider only two technology options: Base load plants and peak load plants.

The costs for the two types are:

$$C_B = k_B P_B + c_B W_B \qquad (6.26)$$

$$C_P = k_P P_P + c_P W_P \qquad (6.27)$$

where index $_B$ and $_P$ represent the base load, typically a coal-fired steam plant, and peak load, typically a gas turbine.

The variables are:

Optimal mix of generating units.

C_P — total cost for peak load units. [NOK]
P_P — total installed capacity for peaking units. [kW]
k_P — fixed costs for peaking units. [NOK/kW year]
W_P — produced electricity in the peaking units. [kWh]
c_P — marginal generation cost (STMC) for the peaking units. [NOK/kWh]

and correspondingly for the base load units.

We assume that there is a large number of standard units so the total capacity of the two types of units can be regarded as continuously variable. We are looking for a minimum cost solution, i.e. a solution where the sum of variable and fixed costs is minimized. The only constraint to the problem is that maximum load should be covered, which means that $P_B + P_P$ should equal maximum load plus required reserves. One implication of this is that if a certain amount of base load capacity is added, the same amount of peak load capacity must be removed in order to keep the capacity balance unchanged.

In order to find a first order condition for this cost minimum, we introduce a small addition to the base load capacity and a corresponding reduction in the peak load capacity:

$$P'_B = P_B + \Delta P \qquad (6.28)$$

$$P'_P = P_P - \Delta P \qquad (6.29)$$

where ΔP is the marginal capacity change. The corresponding change in generated energy will be $\Delta P \cdot T$, which will be added to the base load generation and subtracted from the peak load generation. The new quantities will be:

$$W'_B = W_B + \Delta P \cdot T \qquad (6.30)$$

$$W'_P = W_P - \Delta P \cdot T \qquad (6.31)$$

Where T is the duration time indicated in Figure 6.33. It equals the maximum duration for the peaking units.

The first order condition for a cost minimum is that this small "swap" of capacity has no net impact on the total cost:

$$C_P + C_B - [C'_P + C'_B] = 0 \qquad (6.32)$$

By using (6.30) and (6.31) that leads to:

$$P_P k_P + W_P c_P + P_B k_B + W_B c_B - \\ [(P_P - \Delta P)k_P + (W_P - \Delta PT)c_P + (P_B + \Delta P)k_B + (W_B + \Delta PT)c_B] = 0 \qquad (6.33)$$

which can be easily reduced to:

$$k_P + Tc_P = k_B + Tc_B \qquad (6.34)$$

Equation (6.34) corresponds to the cross point in the lower part of Figure 6.33Figure 6.33What is shown in the figure is that a peaking unit has the lowest cost up to a utilization time of T. With a longer utilization time, base load units will produce more cheaply.

We can solve (6.34) with respect to T and get:

$$T = \frac{k_B - k_P}{c_P - c_B} \qquad (6.35)$$

So the optimal mix of generating capacity can be found simply by calculating T from Equation (6.35) and then finding P_P and P_B from the load duration curve as shown in Figure 6.33 over

6.7.2 The revenue problem

If we now anticipate that the price of electricity equals SRMC at each instant of time either because it is the utility's pricing policy or because the electricity is sold on an efficient competitive market, the consequence will be that in time span 0 to T, the price will be c_P and in the time span T until 8760 h the price will be c_B. That gives the following income:

$$I = (W_P + P_B T)c_P + (W_B - P_B T)c_B \qquad (6.36)$$

We include T from Equation (6.35) and get:

$$I = W_P c_P + W_B c_B + (c_P - c_B) P_B \frac{k_B - k_P}{c_P - c_B} \qquad (6.37)$$

which leads to:

$$I = W_P c_P + W_B c_B + P_B (k_B - k_P) \qquad (6.38)$$

The two first elements on the right-hand side of Equation (6.38) represent the income needed to cover the operation costs of peaking and base load units. If we exclude these elements, we are left with the operational surplus that is supposed to cover the capital cost of the units:

$$DB_B = P_B (k_B - k_P) \qquad (6.39)$$

We can easily see that this is insufficient to cover the fixed costs (the capital costs).

If we have "separate accounts" for the peaking and base load units, we will get the following capital remuneration for the peaking units:

$$DB_P - 0 \qquad (6.40)$$

and for the base load units

$$DB_B = P_B (k_B - k_P) \qquad (6.41)$$

That means that there will be insufficient revenue to cover the capital cost of the base load units and there will be no operational surplus at all to cover capital costs for the peak units.

Based on these simplified assumptions, it is obvious that full remuneration of capital costs will be a problem under a pricing regime based on SRMC. (We return to this discussion in more depth later. A key factor in this discussion is demand side elasticity. We have assumed zero price elasticity here, which is clearly unrealistic). We should also take into account that there is a need for additional (peaking) capacity for reserve purposes. This capacity margin is necessary in order to cover incidental outages and unforeseen load variations.

The question is whether it is still possible to create the right incentives for investments in different types of capacity?

6.7.3 Introduction of load shedding

In order to find a solution to this problem, we introduce the assumption that the total capacity is insufficient to cover the peak load all the time. That means that a few hours during peak load, there has to be some load shedding. The price reflecting the value of lost load (VLL), under such circumstances is very high. According to our assumptions, this price is supposed to penetrate the whole market.

Under these circumstances, the capacity mix will be as indicated in Figure 6.34. The price will be as indicated in the bottom part of the figure.

The optimal mix of generating capacity can be found based on an analogy with Equation (6.35). The optimal mix is given by:

$$T_1 = \frac{k_P}{VLL - c_P} \quad (6.42)$$

(it is assumed to be no investments connected to the load shedding) and

$$T_2 = \frac{k_B - k_P}{c_P - c_B} \quad (6.43)$$

We can now find the operation surplus for the peaking and base load units given this optimal mix.

For the peaking units there will be an operation surplus in hours with a price equal to VLL:

$$DB_P = (VLL - c_P) P_P T_1 \quad (6.44)$$

We include T_1 from Equation (6.42) and find that:

$$DB_P = P_P k_P \quad (6.45)$$

which means that the operation surplus exactly covers the capital cost.

Optimal mix of generating units.

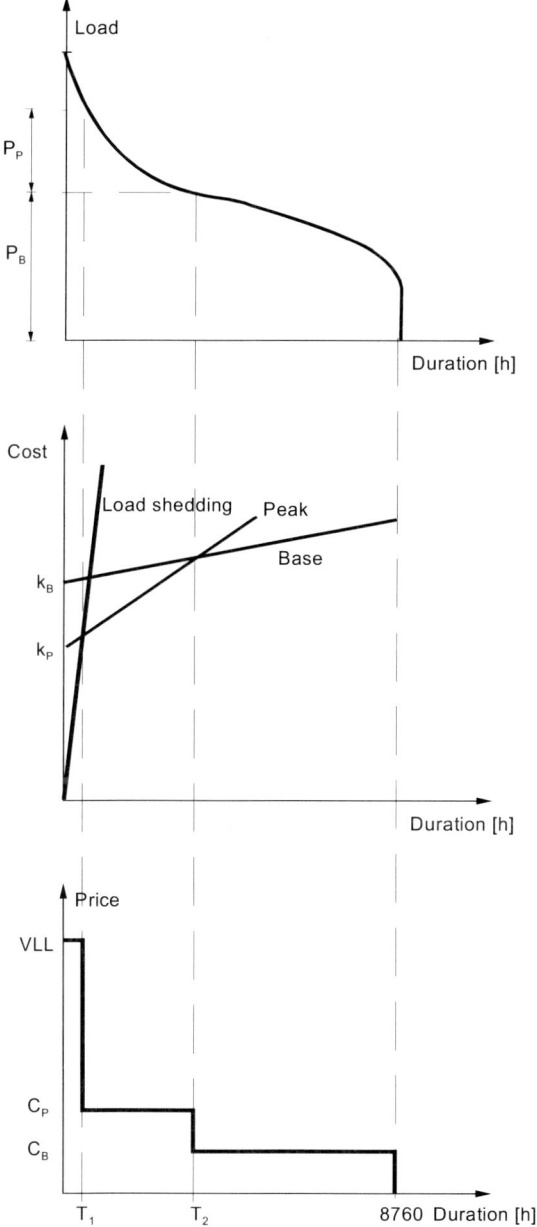

Figure 6.34 Optimum mix of capacity with load shedding during peak.

For the base load units there will be an operational surplus in two periods. In the period 0 to T_1 where the price is VLL and in the period $(T_1 - T_2)$ where the price is c_P. This gives the following surplus:

$$DB_B = (VLL - c_B) P_B T_1 + (c_P - c_B) P_B (T_2 - T_1) \quad (6.46)$$

which can be reformulated to:

$$DB_B = [(VLL - c_P) T_1 + (c_P - c_B) T_2] P_B \quad (6.47)$$

Substituting for T_1 and T_2 from Equation (6.42) and (6.43) we get:

$$DB_B = [\frac{(VLL - c_P) k_P}{VLL - c_P} + \frac{(c_P - c_B)(k_B - k_P)}{c_P - c_B}] P_B \quad (6.48)$$

which gives:

$$DB_B = k_B P_B \quad (6.49)$$

So the conclusion is that the operation surplus covers the capital costs exactly in this case too.

The assumptions on which this analysis is based are by no means realistic. We assume that the capacity is fully utilized under peak load, which means that exactly the right amount of load shedding must be done. At the same time we must avoid uncontrolled breakdown. It is also assumed that curtailment cost (VLL) penetrate the whole market. This will normally not be the case. Despite these unrealistic assumptions, the analysis gives a clue about how the capacity problem can be treated.

6.8 References

[1] A. J. Wood, B. F. Wollenberg: "Power Generation Operation and Control", John Wiley & Sons 1996.

[2] E. S. Huse: "Power generation scheduling". PhD Thesis, Department of Electrical Power Engineering NTNU, 1998.

References

[3] T. Juell Larsen: "Daily Scheduling of Thermal Power Production in a Deregulated Electricity Market". PhD Thesis, Department of Electrical Power Engineering, NTNU, 2001.

[4] B. K. Edwards: "The Economics of Hydroelectric Power". Edgard Elgar publishing, 2003.

[5] G. Doorman: Hydro Power Scheduling, Lecture note Department of Electric Powe Engineering, NTNU, 2009.

[6] V. Hveding: "Digital simulation techniques in power system planning". Economics of Planning, 1968.

[7] A. Haugstad, O. Rismark: "Price Forecasting in an Open Electricity Market based on System Simulation". Proceedings form EPSOM'98, Zurich, Switzerland, September 1998.

[8] A. Gjelsvik, M. M. Belsnes, A. Haugstad: "An algorithm for stochastic medium-term hydrothermal scheduling under spot price uncertainty". Proceedings 13th Power Systems Computation Conference, Trondheim, Norway, June28-July 2nd, 1999, pp. 1079-1085.

[9] B. Mo, A. Gjelsvik, A. Grundt, K: Kåresen: "Optimization of Hydropower Operation in a Liberalized Market with Focus on Price Modelling". Power Tech 2001, Porto, May 2001.

[10] Finn R. Førsund :"Hydropower Economics". Springer, 2007.

[11] Fosso, A. Haugstad, B.Mo and I. Wangensteen: "Generation Scheduling in a Deregulated System: The Norwegian Case". IEEE Trans. on Power Systems, vol. 14, pp. 75-81, February 1999.

[12] O. Wolfgang, A. Haugstad, B. Mo, A. Gjeldsvik, I. Wangensteen, G. Doorman: "Hydro reservoir handling in Norway before and after deregulation." Energy 34 (2009) 1642 - 1651

7 Grid access

7.1 Introduction/definition of concepts

Within the traditional structure of the power supply industry, generation, transmission and distribution were more or less integrated, vertically as well as horizontally. In some cases there was complete vertical integration comprising generation, transmission and distribution. In other cases generation and transmission were integrated, while distribution was left to separate companies.

Within this traditional monopolistic structure there were arrangements for transit of power ("wheeling"). In some cases two power supply companies could use a third party's grid, or a transmission line belonging to one or both of the involved companies, for the transmission or exchange of power. Before the restructuring, this type of arrangement was used for power exchange between the Nordic countries (further discussed in Section 7.5). The objective was normally to reduce the cost and increase the security of supply. Cost minimization was achieved by always using the cheapest available generating unit (correction made for losses and constraints on the transmission). Decisions on how the transmission lines should be used were taken directly by the dispatching centres on a cost minimization basis. How the benefit was divided between the parties was subject to negotiation. In addition to cost minimization, these arrangements contributed to improved security of supply.

Transit arrangements between vertically integrated utilities might involve the use of a transmission system owned by a third party (therefore the term: Third Party Access (TPA)). Compensation to this third party could be negotiated on a case-by-case basis (Negotiated Third Party Access, NTPA), or it could be based on pre-calculated tariffs. This type of transit arrangement between two (vertically integrated) utilities over a third utility's grid, could have a favourable impact on operational costs. However, it did not directly impact consumer prices or in any fundamental manner affect the way scheduling and dispatch were performed. Due to the reduced costs it could still have an influence on consumer prices, but the effect was limited.

As described in Chapter 5, the restructuring of the power system includes a division or unbundling of the system. The supply system was divided in two: one competitive part (generation) and one non-competitive, i.e. monopolistic part represented by the transmission and distribution (T/D) grid. In addition the electricity consuming entities, which also represent a competitive element in the power system, are divided from the rest of the system (as they have always been).

Grid access is vital to the competitive parties. Producers and consumers must be able to use the grid for transportation and it represents a physical marketplace where trade can take place.

Third party access (TPA) is a necessary condition for an efficient market. However, to what extent an efficient market is achieved, is also dependent on the transmission and distribution (T/D) tariffs.

Sections 7.2 and 7.3 deal with the problem of how to use transmission pricing as a means to obtain optimal dispatch. That means it is limited to short-term operational decisions. Transmission pricing will also affect the long-term operational and investment decisions, but that aspect is not discussed here.

We concentrate on how prices affect the grid users and less attention is given to how transmission tariffs affect the grid company. We assume that the grid owner or the System Operator (ISO), who is responsible for transmission pricing, has an objective of maximizing the benefit to society and is using prices to achieve that goal.

In a system with traditional vertically integrated utilities, pricing (or more precisely: tariffing) is a matter between the utility and the end users. Optimal operation of the system including generation scheduling and dispatch is an internal problem for the utility. A dispatching centre is normally responsible for that. In this context, pricing is not interesting for the generators, but it is interesting as a steering instrument towards the consumer. In order to obtain not only an optimal generation dispatch, but an overall societal optimum in which both supply and demand are included, the consumers should see correct prices, i.e. prices reflecting Short-Term Marginal Cost (STMC).

Introduction/definition of concepts

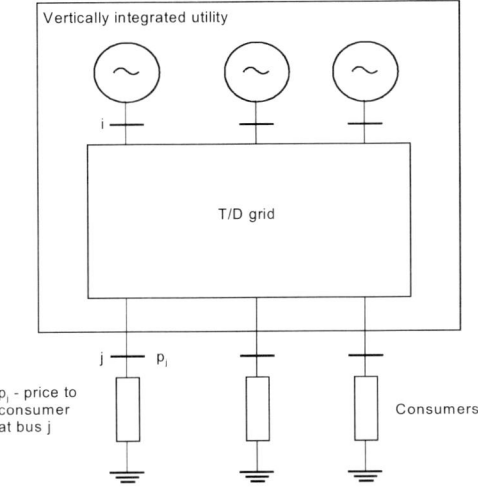

Figure 7.1 Vertically integrated utility.

A power transmission grid is a complex transportation system. The flow is governed by the physical laws of Kirchhoff and Ohm. Generally any change of input/output in one part of the grid will impact the flow in the whole grid. Normally the flow cannot be controlled by the operator,[1] and in a meshed network the flow of electricity follows routes that are not always optimal from an economic point of view. The flow is optimal with respect to physical losses (minimum losses for given input/output).

The problem is that the action from one entity using the grid affects other entities' costs or possibilities of using the same grid. In economic terms that means externalities. A way of solving the problem of *externalities* is to include the external costs in the price in such a way that the price reflects the true cost imposed on the system.

If generation and transmission/distribution (T/D) are split into separate commercial units (*unbundling*), pricing will serve as a signal to generators as well as consumers. A single buyer model[2] (SBM) is based on the notion that the T/D

---

[1] There are certain possibilities. A DC-connection can be controlled. So-called FACTS-devices can be used to control the flow in an AC-network.

[2] The single buyer model (SBM) was discussed some years ago in the EU following a French proposal. The assumption was that the single buyer is normally an integrated company and the independent power producers (IPPs) would have to compete with the single buyer's own generating capacity. The SBM seems to gain less interest today. One interesting application of the single buyer principle is its use for *reactive power*. According to present regulations in

company has full control over the prices on both sides, see Figure 7.2. It is buying from generators on the one side and selling to consumers on the other. The prices can be set with the same optimality objective as in the vertically integrated utility. (We assume that socio-economic optimality is the objective of the grid company.) Prices to the consumers, p_j, can be the same as in the vertically integrated case and prices to the generators, p_i, should be set to obtain an optimal generation dispatch.

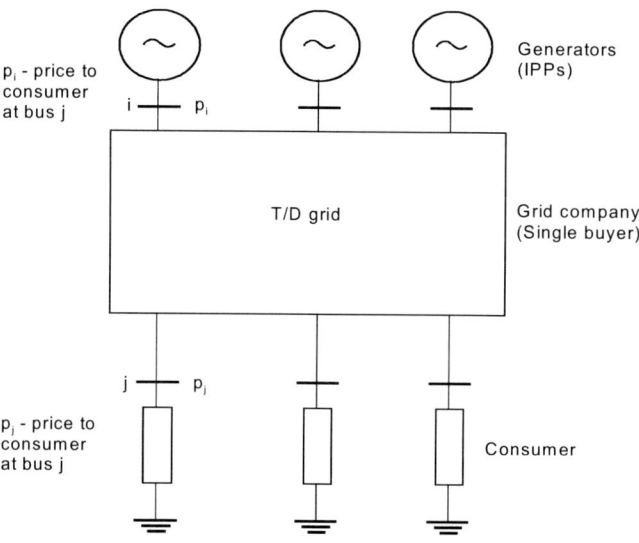

Figure 7.2 Single buyer model.

To some extent a single-buyer model will open up for competition between IPPs. Still it is based on centrally controlled prices, which is not in line with free market conditions.

In order to obtain a system resembling more closely a competitive free market, the grid can be opened for general *third party access* (TPA) (or simply: open access), which means that generators and consumers can get in touch and engage in direct trading transactions. The T/D company offers transportation services and is only responsible for T/D tariffs. The electricity price (market price) is settled through the normal supply/demand mechanism.

Norway, the grid companies have an exclusive right to buy and sell reactive power, which in fact make them "single buyers" for reactive power.

Introduction/definition of concepts

Figure 7.3 illustrates open access with transmission tariffs a_{ij} for transmission between point i and point j. Such *point-to-point tariffs* for transfer are suited for bilateral trading transactions. It is normally up to the trading partners how the transmission charge should be shared between the two.

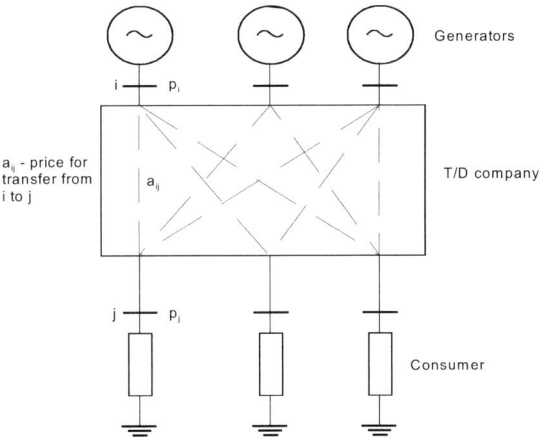

Figure 7.3 Open access. Point-to-point transfer tariffs.

But a power exchange or a pool cannot be combined with a point-to-point tariff system. If one sells through a pool, it is not known who is on the receiving end[3]. A *point tariff* (or *point-of-connection-tariff*) is suited for that type of arrangement. With a point tariff system, the tariff is attributed to the point of connection. What each participant is paying, depends on where the point of connection is, not the connection point of the other trading party. Transmission pricing based on a point tariff also makes bilateral market transactions easier. It is easier for a consumer to compare offers from different suppliers. It is easier to switch from one supplier to another.

A point tariff system requires a *market place* or a *hub*[4] somewhere in the grid. It is not necessarily a fixed bus, and it does not even have to be explicitly defined as a physical location. The point is that every market participant will see a market price - it can be a spot price or a contract price - and a price for transfer between the

---

[3] This is the case in Nord Pool but not necessarily in all power exchanges. It can be based on bilateral trading arrangements, which is the case with the new trading system in England, NETA.

[4] The term "hub" is corresponding to what is called a *hub-and-spoke* tariff system. It is in principle the same as what is call here a *point tariff* in line with established Nordic terminology. (NO: punkttariff)

Chapter 7: Grid access

market place and the local bus where the participant is connected. It is this local price, that is interesting for the participants. This local price is:

- for a consumer: market price + point tariff
- for a generator: market price - point tariff.

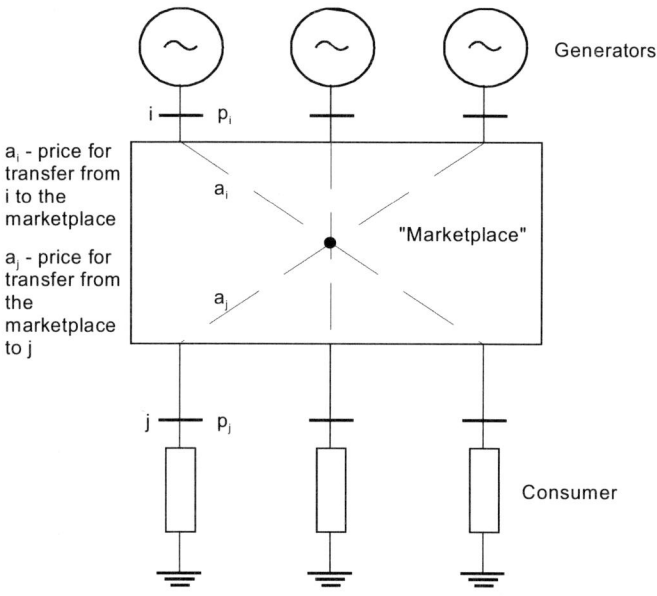

Figure 7.4 Open access. Point tariff system.

This chapter next discusses congestion management based on a single transmission line connecting two nodes (Section 7.2.). Then Section 7.3 introduces a mashed network where we have losses in addition to congestions. Finally, in Section 7.4, there is a description of practical implementation of point tariffs in the Nordic system.

7.2 Congestion Management

7.2.1 Introduction

Congestion management is here discussed on the basis of a very simple grid example. It is one line (or it can be a set of parallel lines) connecting two nodes.

Congestion Management

The transfer capacity is limited due to thermal constraints or other factors. We disregard transfer losses.

> *Different concepts concerning transfer capacity*
>
> In some cases it can be relevant to distinguish between different concepts concerning transfer capacity. We can start with the *Total Transfer Capacity (TTC)* representing the total physical limit. If that is lowered by the *Transmission Reliability Margin (TRM)*, we get the *Net Transfer Capacity (NTC)*. That can be partly occupied by different kinds of reservations such as *Firm Transmission Rights (FTR)* or *Notified Transmission Flow (NTF)* (the concepts are different but the physical reality can be the same) and we get *Available Transfer Capability (ATC)*. These concepts are important in practical congestion management, but in this chapter we will not distinguish between different types of transfer capacities.

That means there is no need to take the influence of loop flows and impedances into account. We can simply assume that all excess generation in one node will be transferred to the other to replace a corresponding deficit there. The energy is transferred from a low price to a high price area. In a more complex grid with loop-flow possibilities, that will not always be the case.

Two different congestion management procedures are described: One is the *area pricing or zonal pricing*. The other is the so-called *buy-back* procedure.[5]

7.2.2 Congestion Management Based on Area Pricing

We shall look at how the area pricing mechanism is used to control the quantity transferred over a congested interconnection so that the available capacity is fully utilized. The procedure is equal to the one Nord Pool is using in the Nordic system (except for the fact that there are more then two price areas in the Nordic system, which increases the complexity).

We use a two-area example as shown in Figure 7.5.

---

[5] There is presently no established common terminology is for these concepts. The term *area pricing* is used in the Nordic system while *zonal pricing* was used about essentially the same thing in California. *Market splitting* is also used for this procedure. A similar procedure is called *implicit auction*. The term *implicit* congestion management is used in some cases for what we here a call buy-back procedure. This is also called *counter-trade*. As opposed to this implicit procedure, the term *explicit* congestion management is used for area or zonal pricing. Here we are using a terminology in line with what is currently used in the Nordic system.

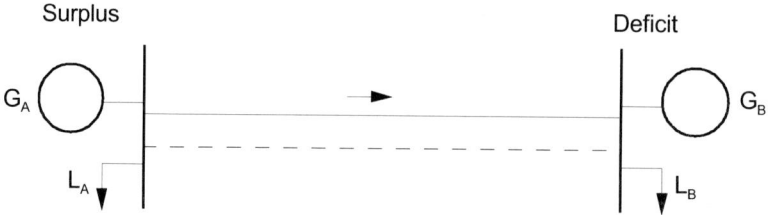

Figure 7.5 Transfer between a surplus and a deficit area

We assume an ideal open market in both areas connected by the line. We have a surplus area A and a deficit area B. If there is sufficient capacity, power will flow from the surplus to the deficit area without significant price differences. That flow will be reduced if the price in A is lowered and/or the price in B is raised. Optimal congestion management means that the prices are adjusted exactly so much that the transferred power matches the available capacity.

If there is congestion between the two areas indicating that there is an active capacity constraint, it will have an impact on the involved parties. In the surplus area (A), the price will be low and in the deficit area (B) the price will be high compared with a non-congested state. The grid company (or the SO or the PX or pooling institution depending on the institutional arrangement) will receive a profit i.e. a congestion rent. Generators in A and consumers in B will be losers while generators in B and consumers in A will be winners; again compared with a non-congested situation. Totally the congestion leads to a loss to society, i.e. a welfare loss.

In order to explain the area pricing procedure in more detail, we introduce supply and demand curves[6] to describe the market in the two areas. We assume simple piecewise linear curves in order to get a simple numerical example.

---

[6] We assume the classical (Marshallian) supply and demand curves here. These are based on the notion that the demand curve stems from the consumers and the supply curve stems from the producers. That is generally not the case with the Nord Pool price cross. Behind Nord Pool bids there can be a mix generators and consumers on the selling side as well as the purchasing side. The conclusions from our discussion here are still valid.

Congestion Management

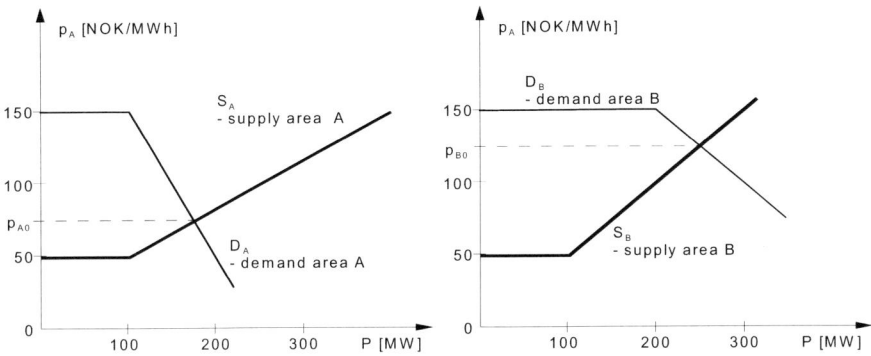

Figure 7.6 Market balance in a surplus (low price) area (A) and deficit (high price) area (B).

We start with a situation where there is no connection between the two areas, see Figure 7.6.

The market balances are as indicated in the figure. Prices and quantities are:

p_A = 75 NOK/MWh P_A = 175 MW p_B = 125 NOK/MWh P_B = 250 MW

We can find the producer and consumer surplus for the two areas by simple integration. The producer surplus in A is the area between the price line the supply curve:

$$\text{Producer surplus in } A = \int_0^{P_{A0}} (p_{A0} - S_A(P)) dP \quad (7.1)$$

The consumer surplus in A is the area between the demand curve and the price line:

$$\text{Consumer surplus in } A = \int_0^{P_{A0}} (D_A(P) - p_{A0}) dP \quad (7.2)$$

The producer and consumer surplus in B can be calculated in the same manner.

With the example shown in, we can easily find these quantities:

185

Chapter 7: Grid access

Consumer surplus in A	NOK	10 312.50
Producer surplus in A	NOK	3 437.50
Consumer surplus in B	NOK	5 625. -
Producer surplus in B	NOK	13 125. -
Total surplus	NOK	32 500. -

Next we assume there is an interconnection without transfer restrictions between the two areas. It can then be regarded as one integrated market and we can find total supply and total demand by adding supply in A to supply in B (so-called "horizontal addition" see Chapter 2) and the same for the demand curves in the two areas. The total supply and demand curves will then be as shown in Figure 7.7.

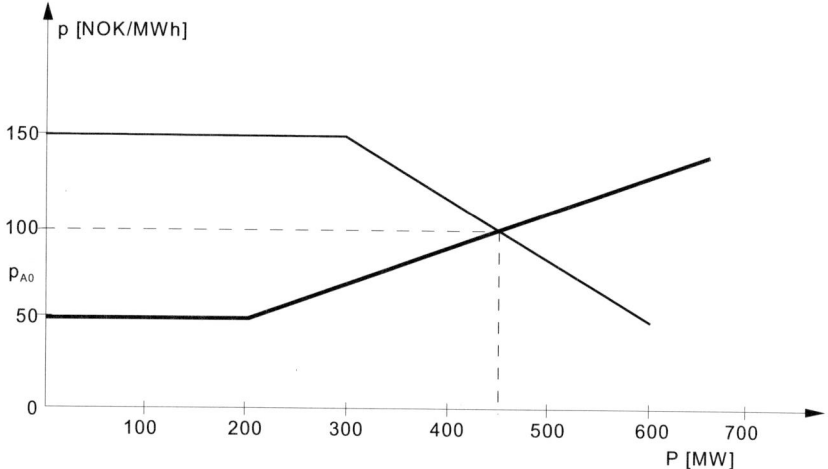

Figure 7.7 Aggregate supply and demand curves.

The price (the System Price in Nord Pool's terminology) is now 100 NOK/MWh and the quantity is 450 MW. We can find the producer and consumer surplus in the same manner as before. We find that:

Consumer surplus for the integrated area:	NOK	18 750
Producer surplus for the integrated area:	NOK	16 250
Total surplus for the integrated area:	NOK	35 000

We notice that the total surplus for the integrated area is higher than the total surplus for the two areas if they are split. There is an additional surplus of NOK 2500 for the integrated area. That represents the economic benefit to society of having sufficient transfer capacity between the two areas.

What is sufficient capacity in this case? If we go back to Figure 7.6, we can find the capacity needed to give a price of 100 NOK/MWh in both areas. First we draw a horizontal line representing the common price of 100 NOK/MWh through both diagrams. This is shown in Figure 7.8. Now we can easily see that this price gives a capacity surplus in area A of 100 MW and a corresponding deficit in area B. So we can conclude that it is necessary to have a transfer capacity of 100 MW to give equal prices in both areas.

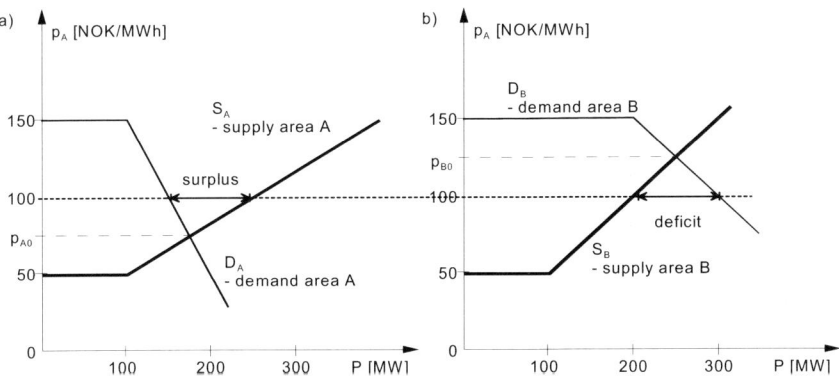

Figure 7.8 Surplus and deficit in two areas with equal prices.

If the transfer capacity is less than 100 MW, there will be a price difference between the two areas. In order to study the consequence of such a limited capacity, we introduce a transfer capacity of 50 MW. Figure 7.9 shows the consequence of that. We start with zero capacity at the top a), there is 50 MW in the middle b) and there is full capacity i.e.100 MW at the bottom c).

Introduction of transfer capacity can be looked on as a shift of the demand curves. If we compare Figure 7.9 a) and b), we see that the 50 MW transfer capacity is equivalent to a 50 MW shift of the demand curve to the right in area A. A connection to the high price area (B) represents an increased demand seen from the generating companies in A. The opposite is the case in area B. The fact that 50 MW of the demand in B is covered with transfer from A, implies a corresponding decrease of the demand (in B) that has to be covered by generation in B. That implies a shift of the demand curve to the left.[7] The new price in A will be NOK 87.5 (compared to the former NOK 75) and the new price in B will be NOK 112.5 (compared with the former NOK 125).

This introduction of transfer capacity affects the producer and consumer surplus on both sides. Again we can compare Figure 7.9 a) and b). Starting with area A, we see that consumer surplus decrease from a) to b) (less is consumed at a higher price) and producer surplus increase (more is sold at a higher price). We also see that producers gain more than the consumers lose, giving a net benefit to society. This is represented in Figure 7.9 b) by the small triangle marked NBA. (Net Benefit area A).

The impact on consumers and producers in area B is the opposite of what we saw in A. When we compare a) and b), we see that the producer surplus in area B decreases and consumer surplus increases with the introduction of transfer capacity. It is exactly parallel to the development in A, but with the opposite sign. In this case we also obtain a net social benefit, represented by area NBB in figure b) (Net Benefit area B).

NBA and NBB are net benefits for the market participants, i.e. consumers and producers, on both sides. In addition there is also a trading surplus because of the transfer of electricity from a low price to a high price area. This trading surplus, normally called a *congestion rent,* can go to the System Operator, the Power Exchange or the grid company.[8] The congestion rent equals the transfer capacity (which equals the actual transfer over the congested interface if the grid is optimally utilized) multiplied by the price difference between the two sides. This congestion rent is also part of the overall economic benefit of establishing a transfer capacity between the two areas.

The total economic benefit of establishing this interconnection of 50 MW as indicated in b), will be:

---

[7] We have used shift of demand curves here to illustrate the effect of introducing increasing transfer capacity between the two areas. That is not the only way to do it. A shift of the supply curves (in the opposite direction) can be used as an alternative. The alternative procedures lead to the same conclusions.

[8] In the Nordic system this congestion rent, which is originally collected by Nord Pool, goes to the TSO. In case different TSOs are affected, the congestion rent is split between them. To what extent the TSOs keep the congestion rent, depend on the economic regulation. With the present revenue regulation in Norway, the TSO (Statnett) will not benefit from the congestion rent.

NBA:	50 · 12,50 · ½	= 312,50 NOK
NBB:	50 · 12,50 · ½	= 312,50 NOK
Congestion rent:	50 · 25,00	= 1250,00 NOK
Total:		1875,00 NOK

If we go one step further and increase the transfer capacity to 100 MW, the consequences will be as shown c). The price will now be equal on both sides: 100 NOK/MWh, which is in line with former conclusions. The benefit for the market participants will be as indicated in the figure (NBA and NBB). In this case there will be no congestion rent because there is no price difference. So the total economic benefit of this 100 MW connection will be:

NBA:	100 · 25 · ½	= 1250,00 NOK
NBB:	100 · 25 · ½	= 1250,00 NOK
Total:		2500,00 NOK

This is also in line with former conclusions.

Chapter 7: Grid access

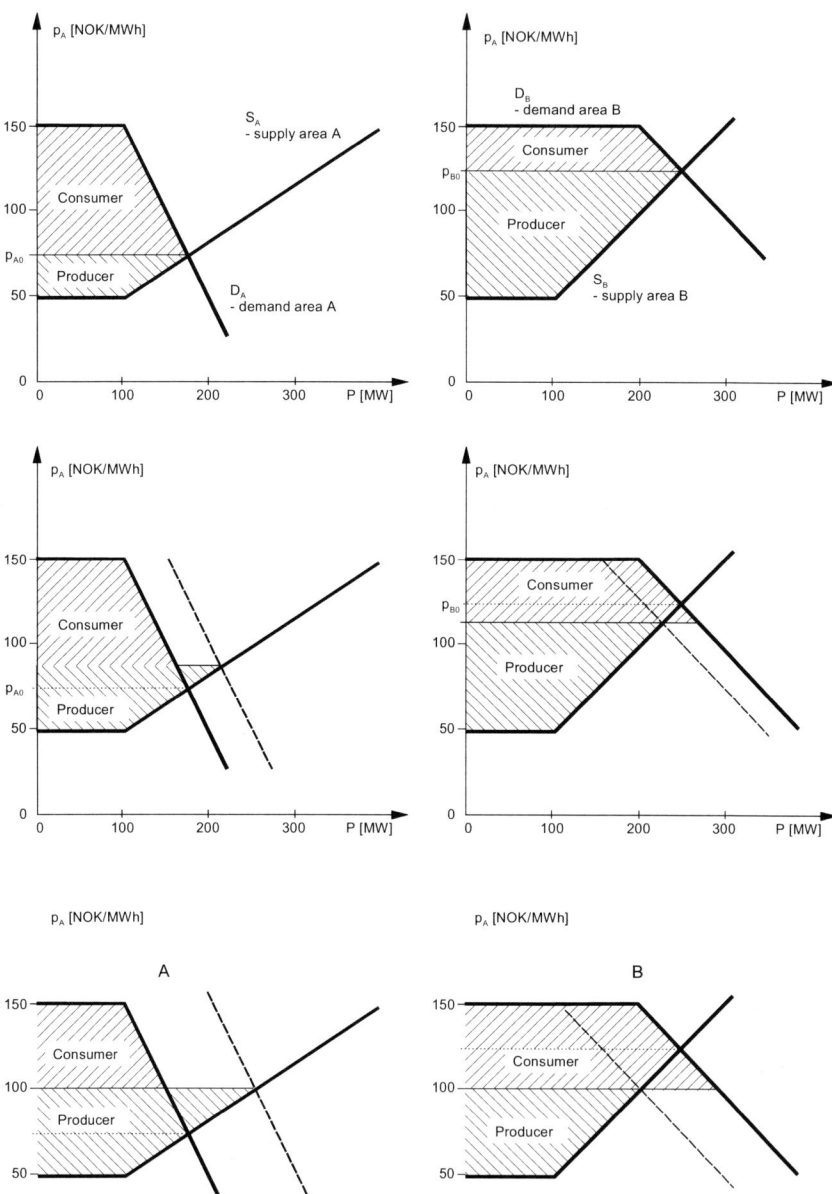

Figure 7.9 Calculation of producer and consumer surplus under different transfer restrictions. The starting point is a high price in B and low price in A.

Table 7.1 Economic consequences of increased transfer capacity [NOK/h]

Transfer capacity (MW)	0	50	100
Consumer surplus A	0	-2109.375	-4062.50
Producer surplus A	0	+2421.875	+5312.50
Consumer surplus B	0	+3281.250	+6875.00
Producer surplus B	0	-2968.750	-5625.00
Net benefit to consumers and producers	0	625.000	2500.00
Congestion rent	0	1250.000	0
Total economic benefit	0	1875.000	2500.00

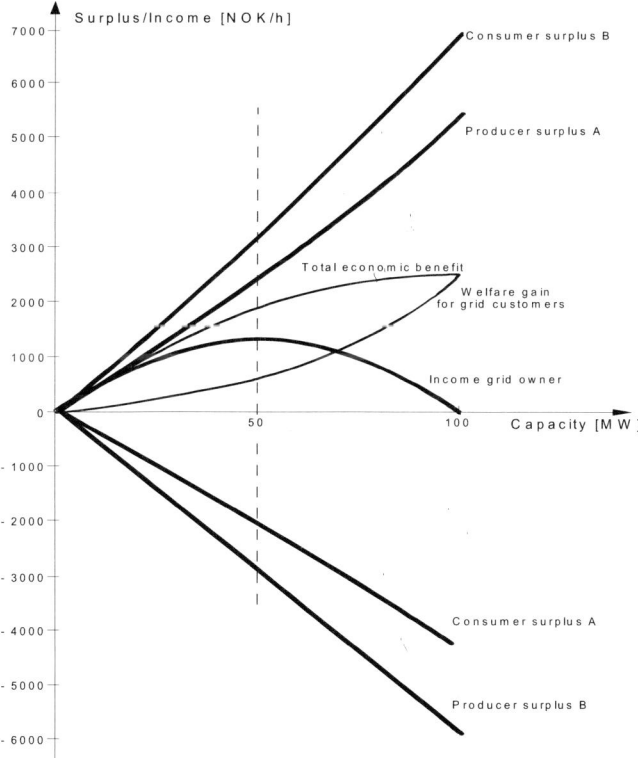

Figure 7.10 Economic consequences of increasing transfer capacity between the two areas.

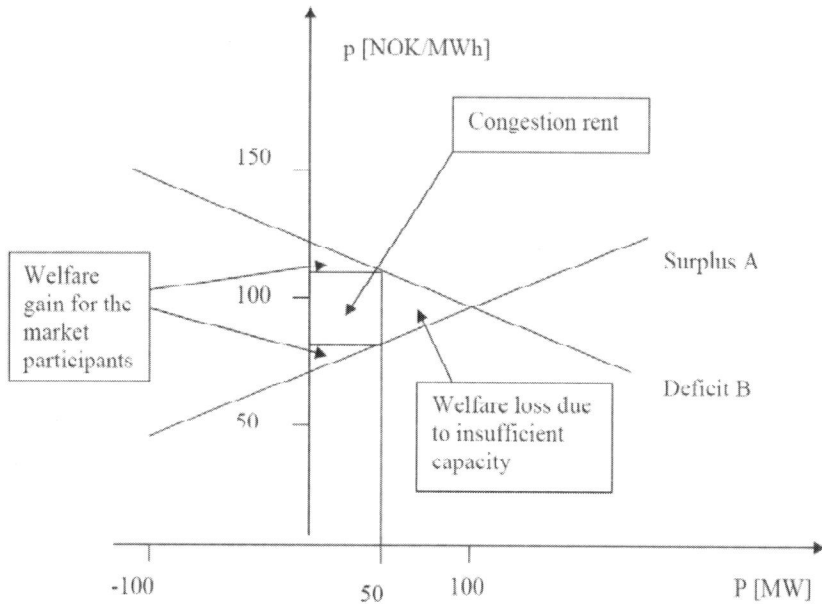

Figure 7.11 Illustration of 50 MW transfer capacity between surplus area A and deficit area B

A simplified illustration of the consequence of having limited transfer capacity is shown in Figure 7.11. The example is the same as before. The line representing net surplus in area A is found by subtracting supply from demand in area A:

$$Surplus\ A = (S_A - D_A)$$

The deficit in B is found analogously by subtracting demand from supply in B:

$$Deficit\ B = (D_B - S_B)$$

We assume that we have 50 MW transfer capacity available, while the total requirement is 100 MW. From this illustration we can now see:

- The congestion rent represented by the square, which is the difference between the area prices multiplied by the capacity.

- The welfare gain for the market participants is represented by the two small triangles. This welfare gain plus the congestion rent is the total benefit to society of having this capacity available.
- The triangle to the right represents the loss to society due to insufficient capacity, i.e. the capacity is less than the 100 MW required to get equal prices in the two areas.

By using this simplified illustration instead of the more complex one in Figure 7.9, we lose some information. Information about net gains and losses in both areas is maintained, but we lose information about the division between consumers and producers.

If we look on the economic consequences for the involved parties, it is evident that nobody will receive correct incentives to make investments in transfer capacity. A correct incentive to invest should reflect the economic benefit to society (welfare gain) as a result of increased capacity. The grid company's income will not give the right incentive. The change of income due to increased transfer capacity can even be negative. Producers in A will receive a strong incentive to invest. Under the assumptions made here it will be stronger than what is correct from a societal point of view. A producer in A will benefit from increased capacity, not only through better overall utilization of resources, but even more from a redistribution of economic benefit from the consuming side to the producing side. In principle, the same is the case with consumers in B, but in this case the consumers will benefit at the cost of generators if a grid investment is made. In both cases the redistribution effect is much larger than the small net benefit that. A possible coalition between parties on both sides will further strengthen this over-incentive to invest[9]

If it had been possible to divide the transfer capacity into marginal units, it could have been possible to find a rational solution to the investment problem within this framework. If the grid company could invest in small additions to the transfer capacity - so small that it would have marginal impact on the prices, increased income to the grid company would be the same as the increased social benefit. But this is not possible. Substantial capacity has to be built in one or a few operations.

---

[9] This effort to affect the market price by investing more or less must be regarded as exercise of market power. We will not discuss market power here. It is often argued that the grid company will take revenue effects into account when an investment over a congested interface is considered. If the overall objective for the grid company is to maximize social benefit, that will not be the case. We should also keep in mind that investment incentives are affected by the economic regulation of the grid companies. With the current revenue regulation in Norway, the rent does not contribute to increased net revenue for the grid company. So there is no incentive (or disincentive) to invest due to the congestion rent.

So we are faced with the problem of indivisibility or lumpiness of investments. Investments in the grid can never be marginal.

7.2.3 Consequences of a Buy-Back Procedure

The buy-back procedure means active involvement from the System Operator's side in the sense that it will be engaged in active market transactions. In order to relieve a potentially overloaded interface, the SO buys on the one side and sells on the other. The SO will buy (i.e. pay generators to increase output and end users to decrease consumption) on the deficit side and sell (i.e. pay generators to decrease output and end users to increase consumption) on the surplus side. This implies that the SO will buy at a higher price than it sells and thereby suffer a net loss on the transaction.

Figure 7.12 gives an illustration. The case is the same as in the previous example. We start with the assumption that all the market participants being used for buy-back are paid the same price on the same side of the congested interface. We start with 100 MW transfer capacity and equal prices on both sides. If we now have only 50 MW available, the System Operator must buy in area B (high price) and sell in area A (low price) in order to keep the transferred power within the available limits. We assume the System Operator can buy and sell both on the consumer and the generator sides[10]. It buys production increase (P_{SB} = 25 MW) and consumption decrease (P_{DB} = 25MW) in area B at a price of 112.5 NOK/MWh and sells production decrease (P_{SA} = 37.5 MW) and consumption increase (P_{DA} =12.5 MW) in area A at a price of 87.5 NOK/MWh. The total cost for the System Operator will be 1250 NOK/h while the overall welfare loss is 625 NOK/h.

There are two important differences between area pricing and buy-back that should be noticed:

- Area pricing leads to a net income for the System Operator while buy-back leads to a net cost.

- With buy-back, all market participants (except the ones being directly engaged in the buy-back procedure by the System Operator) will see only one price - the system price - in the market. They will not be confronted with different area prices.

---

[10] That is not always the case. In many cases only the generating side will be involved. The term "off merit scheduling", which is sometimes used for this procedure, refers to this non-optimal rescheduling of generators. This terminology is to some extent linked to old central planning or scheduling procedure. This use of generators only can partly be due to the fact that for different practical reasons, the regulating market is used for this buy-back procedure, not the ordinary spot market. The regulating market has traditionally been a market for generators. But that has changed lately. Today, there is no reason to limit the buy-back procedure to the generating side alone.

There are both advantages and disadvantages for the different solutions tied to these two aspects.

From an investment incentive point of view, it is regarded as an advantage that the System Operator (which in the Nordic system is also the grid operator, a TSO) is exposed to a net cost if there is insufficient grid capacity.

As for the second aspect, the market participants will in general prefer one uniform system price for the whole interconnected area. Different area prices will cause extra uncertainty in trade across potentially congested interfaces. The trading partners run the risk of being confronted with a congestion rent, which can be difficult to forecast and with limited hedging possibilities. The financial Futures and Forward markets of Nord Pool use the system price as reference price, which means it is not possible to use these instruments to hedge against price uncertainties not captured by the system price[11]. Differences between the system price and area prices represent an uncertainty that the market participants prefer to avoid.

On the other hand it is obvious that from a market efficiency point of view, the buy-back procedure does not give an optimal solution. The market participants will not see the correct price in the area they are located. Even if this does not affect short-term adaptation of the participants (which will be the case if the spot market bidding is correct) it can have an impact on the long term behaviour. A potential investor can be influenced by electricity prices in the choice of geographical location.

---

[11] One such hedging instrument, the so-called *Contract for Differences (CfD)*, was recently introduced by Nord Pool. There is limited experience with this market so far.

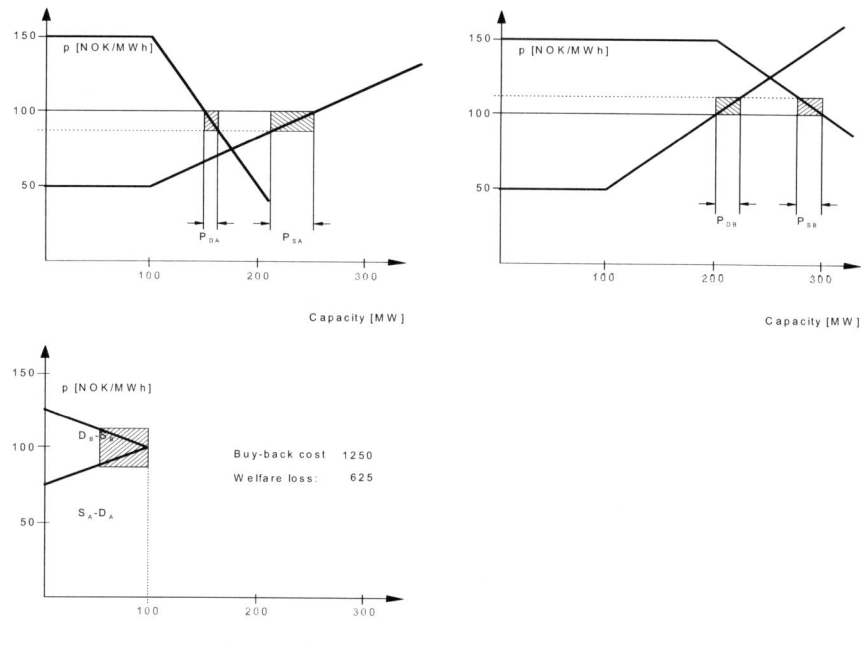

Figure 7.12 Illustration of the buy-back procedure.

The procedure indicated in Figure 7.12 is based on the assumption that all market participants used in the buy-back process, receive the same price. They all receive the market-clearing price. (In this case that is the clearing price in the Regulating Market. For practical reasons it is the Regulating Market and not the Spot Market bids that are used for buy-back, and the figure may be somewhat misleading with respect to this. The Regulating Market will normally have steeper supply and demand curves than the spot bids.)

Another alternative is to use a procedure based on perfect price discrimination. That means all participants being used, receive the price they bid, not the market clearing price[12]. This leads to lower cost for the SO and to a lower average income for the market participants. The procedure is shown in Figure 7.13.

---

[12] This is difficult to obtain in practice. A pay-as-bid mechanism will affect the bidders. They will not reveal their true marginal cost or marginal utility.

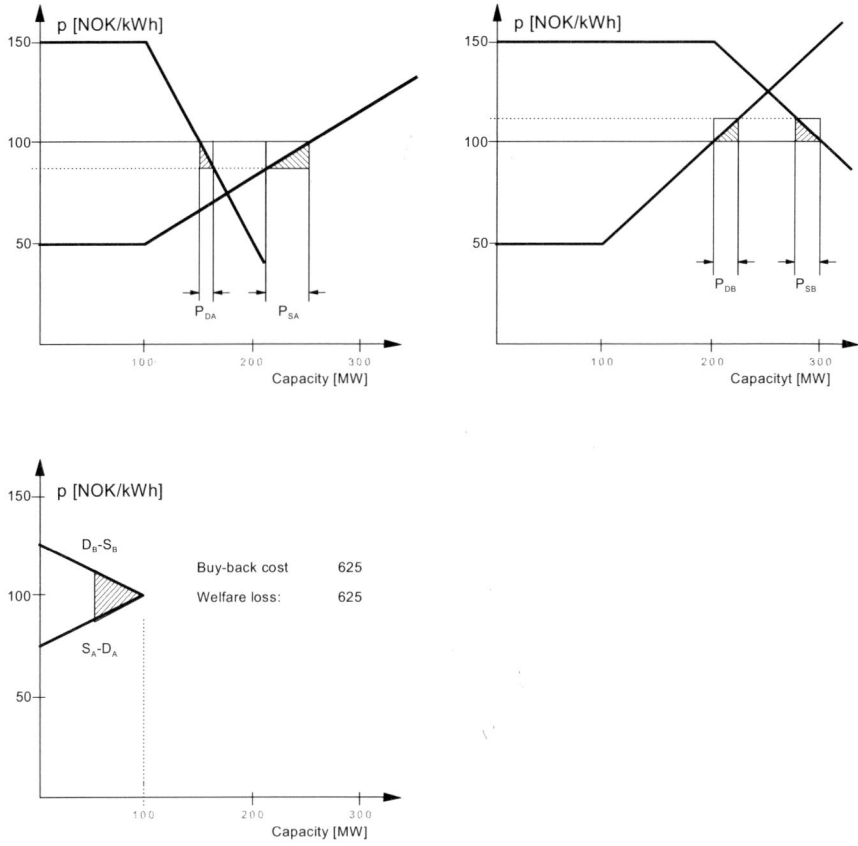

Figure 7.13 Illustration of buy-back procedure with price discrimination.

Figure 7.14 gives a simplified illustration the different congestion management mechanisms.

b) Represents the area pricing

c) Represents the ordinary buy-back

d) Represents buy-back with price discrimination.

It should be noticed that the buy-back procedures used in practice are making use of the regulating power market, not the spot market. So these comparisons are not quite relevant with respect to current practice.

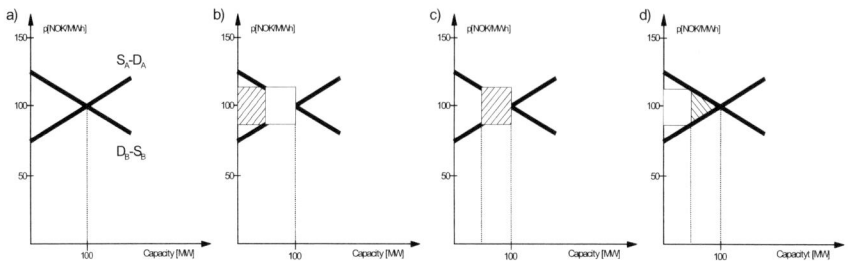

Figure 7.14 Illustration of different congestion management mechanisms

7.3 Price Calculation Based on Optimal Power Flow (OPF)

In this section we show how optimal power flow (OPF) can be used partly as a theoretical basis for a pricing concept, and partly as a practical tool for price calculations.

Ordinary OPF can be extended to include security considerations, so-called security constrained optimal power flow (SECOPF). A simple case is briefly described later. SECOPF is the correct basis for transmission pricing, but security constraints can also be included, although in a simplified way, by adjusting transfer capacity limits with a suitable *Transmission Reliability Margin (TRM)*. In this section we will concentrate on ordinary OPF.

OPF (which is basically a static optimization) is relevant as long as we discuss the optimal dispatch of a thermal system. In a hydro or a hydro-thermal system optimal operation must be seen in a dynamic context due to the importance of intertemporal ties[13]. Transmission pricing in a hydro or hydro-thermal system is discussed in Section 7.3.10.

This chapter contains some examples of OPF or economic dispatch calculations and how optimal transmission tariffs can be derived from that. Different complexities and different types of grid models are used. The number of nodes goes from three to eleven. Generation is based on thermal, hydro or a combination of these.

As for the grid models, there are different alternatives:

- *Non-electrical models*. There are different types. The one used in some of the examples in this section has losses proportional to the square of transferred power.

---

[13] Intertemporal ties are also relevant in a thermal system due to costs and restrictions connected to start/stop, ramping etc. See Chapter 6.

Price Calculation Based on Optimal Power Flow (OPF)

- *DC-models*. These models describe the grid correctly as a DC grid. One advantage of a DC model is that it can have losses as well as congestions.
- *DC-equivalent models*. These models describe an AC grid with a DC-like approximation. A DC-equivalent has no losses, but it can have congestions.
- *Full AC-model*. This describes the grid correctly as an AC grid.

Table 7.2 The examples of grid models that will be used.

System	Grid model
Three node Thermal	DC-model
Three node Hydro	Non-electrical
Three node Hydro/thermal	Non-electrical
11 node Thermal	DC-equivalent
11 node Thermal	AC-model

7.3.1 Principle

An Optimal Power Flow (or Optimal Dispatch)[14] problem can be formulated as follows:

Given a grid with a set of input nodes and a set of output nodes. There are generators connected to the input nodes, each with a given cost function i.e. cost as a function of generation:

$$C_i = C_i(P_{Gi}) \quad (7.3)$$

The marginal cost is:

$$c_i = \frac{\partial C_i}{\partial P_{Gi}} = c_i(P_{Gi}) \quad (7.4)$$

---

[14] The term: Optimal Dispatch or Economic Dispatch is normally used for a minimum generating cost optimization. Optimal Load (or Power) Flow is often based on an extended set of control variables including voltage control, switched capacitor setting etc. Here we are using only power input and output to the grid as control variables.

where:

C_i is the cost of generation in point i.

P_{Gi} is the generation (active power).

c_i is the marginal cost.

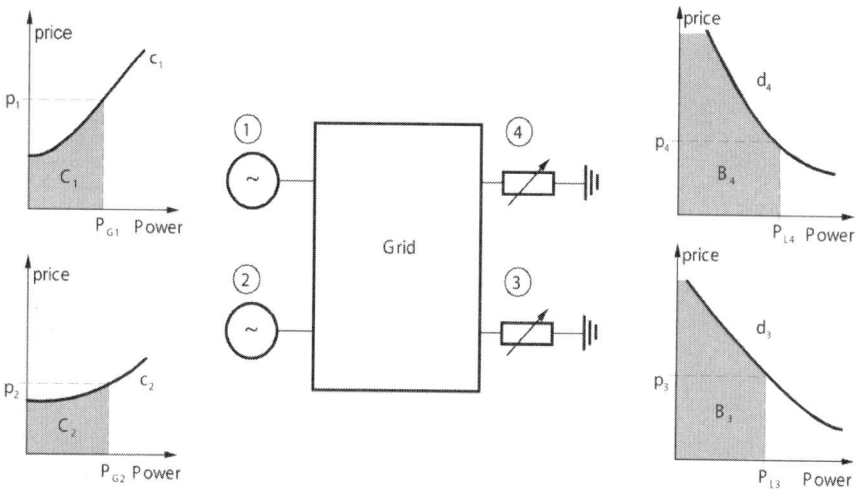

Figure 7.15 Optimal Power Flow.

The consumers are connected to the output nodes, each with a utility function i.e. consumer utility as a function of consumption:

$$B_j = B_j(P_{Lj}) \qquad (7.5)$$

and marginal utility, i.e. marginal willingness to pay:

$$d_j = \frac{\partial B_j}{\partial P_{Lj}} = d_j(P_{Lj}) \qquad (7.6)$$

where: Bj is the total utility for consumer j. P_{Lj} is the electricity consumption of this consumer and d_j is the marginal willingness to pay.

Price Calculation Based on Optimal Power Flow (OPF)

The objective can now be formulated in a conventional way as maximization of utility (or benefit) minus cost:

$$\max(f) = \max(\sum_{j=1}^{m} B_j - \sum_{i=1}^{n} C_i) \qquad (7.7)$$

Where f is the objective function that is maximized with respect to all power inputs and outputs.

It is obvious from this equation that if the consumption is given, which means that the utility is constant (U_j = const.), the objective can be formulated as a cost minimization:

$$\max\left(const - \sum_{i=1}^{n} C_i\right) \Rightarrow \min \sum_{i=1}^{n} C_i \qquad (7.8)$$

The cost (or to be more precise: the variable cost component) equals the area under the marginal cost curve.

$$C_i = \int_{0}^{P_{Gi}} c_i(P)\, dP \qquad (7.9)$$

The utility (or the variable component of the utility) equals the area under the marginal willingness to pay curve.

$$B_j = \int_{0}^{P_{Lj}} d_j(P)\, dP \qquad (7.10)$$

The costs and utilities are illustrated by the shaded areas in

Figure 7.16 and the objective function is the difference between the shaded areas to the right (the utilities) and the shaded areas to the left (the costs). The compulsory restrictions to this optimization problem are the load flow equations. Using vector notations the load flow equations can be written:

$$g(\mathbf{x}, \mathbf{P}) = 0 \qquad (7.11)$$

where **P** is the input/output vector:

$$\mathbf{P} = \begin{bmatrix} P_{G1} \\ \cdot \\ \cdot \\ \cdot \\ P_{Gn} \\ P_{L1} \\ \cdot \\ \cdot \\ \cdot \\ P_{Lm} \end{bmatrix} \qquad (7.12)$$

and **x** is the state vector, i.e. the voltage at every node[15].

$$\mathbf{x} = \begin{bmatrix} U_1 \\ \cdot \\ \cdot \\ \cdot \\ U_{n+m} \end{bmatrix} \qquad (7.13)$$

With the same vector notations the objective function can be written as:

$$f(\mathbf{P}) \qquad (7.14)$$

The extended objective function can now be formulated with the introduction of the Lagrange multipliers,

---

[15] We are not concerned with the detailed formulation of the power flow equations here. For our purpose it is sufficient to know that the multipliers corresponding to the power balances can be found.

Price Calculation Based on Optimal Power Flow (OPF)

$$L(x, P, \lambda) = f(P) + \lambda^t g(x, P) \quad (7.15)$$

where L is the extended objective function (also called Lagrange function), and

$$\lambda = \begin{bmatrix} \lambda_1 \\ \cdot \\ \cdot \\ \cdot \\ \lambda_{n+m} \end{bmatrix} \quad (7.16)$$

are the Lagrange multipliers.

We can now find the first order conditions for an optimum solution by setting the derivative of the extended objective function equal to zero. It can be shown that this leads to the following set of equations.

$$\frac{\partial L}{\partial \lambda} = 0 \rightarrow g(x, P) = 0 \quad (7.17)$$

$$\frac{\partial L}{\partial x} = 0 \Rightarrow \left(\frac{\partial g}{\partial x}\right)^t \lambda = 0 \quad (7.18)$$

$$\frac{\partial L}{\partial P} = 0 \Rightarrow \lambda = \begin{bmatrix} c \\ d \end{bmatrix} \quad (7.19)$$

$$\mathbf{c} = \begin{bmatrix} c_1 \\ \cdot \\ \cdot \\ \cdot \\ c_n \end{bmatrix} \qquad (7.20)$$

are the marginal costs, and

$$\mathbf{d} = \begin{bmatrix} d_1 \\ \cdot \\ \cdot \\ \cdot \\ d_m \end{bmatrix} \qquad (7.21)$$

are the consumers' marginal willingness to pay.

Equation (7.19) applies at buses where generation is not limited by available capacity. (For details on how Equations (7.18) to (7.19) are derived, see the appendix to this chapter).

Equations (7.17) – (7.19) are a set of simultaneous equations that can be solved (provided necessary conditions concerning convexity etc. are met). This is an ordinary unrestricted optimum load flow calculation (i.e. with no restrictions except the load flow equations).

In addition to the load flow equations (7.17) which must always be met, further restrictions can be added. The most relevant are restrictions on transfer capacity, which can written as:

$$\mathbf{h}(\mathbf{x}) \leq 0 \qquad (7.22)$$

and restriction on generating capacity:

$$0 \leq P_i \leq P_{imax} \qquad (7.23)$$

Price Calculation Based on Optimal Power Flow (OPF)

In addition, it is some-times relevant to define restrictions on voltage. Restrictions on voltage will mainly have implications for reactive power. Optimal power flow is an interesting technique for optimal allocation of reactive power, but here we will concentrate on the active power. Hence restrictions on voltage will not be taken into account.

Restrictions on transfer and generating capacity are in many cases binding restrictions and have implications for the solution. These restrictions can be taken into account by introducing the following Lagrange multipliers.

η for restrictions on transfer capacity.
γ for restrictions on generating capacity.

Now we can derive a set of equations for this constrained case. The procedure is the same as in the previous one. We define an extended objective function.

$$L(\mathbf{x},\mathbf{P}) = f(\mathbf{P}) + \lambda^t \mathbf{g}(\mathbf{x},\mathbf{P}) + \eta^t \mathbf{h}(\mathbf{x}) + \gamma^t (\mathbf{P} - \mathbf{P}_{max}) \qquad (7.24)$$

In this case the conditions for optimality are also derived from the Karush-Kuhn-Tucker conditions. This leads to the following set of equations (details in the appendix):

$$\mathbf{g}(\mathbf{x},\mathbf{P}) = 0 \qquad (72.5)$$

$$\mathbf{h}(\mathbf{x}) = 0 \qquad (7.26)$$

$$(\mathbf{P} - \mathbf{P}_{max}) = 0 \qquad (7.27)$$

$$\left(\frac{\partial \mathbf{g}}{\partial \mathbf{x}}\right)^t \lambda + \left(\frac{\partial \mathbf{h}}{\partial \mathbf{x}}\right)^t \eta = 0 \qquad (7.28)$$

$$\lambda = \begin{bmatrix} \mathbf{c} + \gamma \\ \mathbf{d} \end{bmatrix} \qquad (7.29)$$

It should be noticed that in Equation (7.26) h(x) represents active restrictions only, i.e. cases where h(x) = 0. If h(x) < 0, the corresponding equation is not included. The same applies to restrictions on generating capacity, Equation 7.29).

This description of Optimal Power Flow, making use of Lagrange multipliers and the Kuhn-Tucker conditions for optimality, leads to this set of equations (Equations (7.22) to (7.29)). There are different procedures for the solution of this set of equations, and there are methods for solution of the optimization problem not making use of Lagrange multipliers at all. We will not go into the different solution algorithms here. There are several methods available. We will focus on the interpretation of the Lagrange multipliers, especially the lambdas.

7.3.2 Nodal Prices

One advantage of using Lagrange multipliers tied to the power balance in each node is that the multipliers can be interpreted as nodal prices. The lambdas represent the price at each node deriving from the optimal load flow solution. λ_i is the marginal cost of supplying an additional unit of electricity at bus i. This marginal cost is the same if the additional unit is transferred from any other node in the grid or it is provided by local generation or load reduction.

So the nodal prices can be written as:

$$p_i = \lambda_i = c_i + \gamma_i \qquad (7.30)$$

for a generation node, and

$$p_j = \lambda_j = d_j \qquad (7.31)$$

for a consumption node.

η_i is the congestion element representing the local effect of a transfer capacity limit. It is the marginal benefit (improvement of the objective) of additional transmission capacity.

γ_i is the marginal value of generation capacity at node i. If you add a marginal unit of capacity (with marginal generation cost c_i) at node i, its local value will be γ_i.

Before deregulation, optimal power flow was traditionally based on the assumption that there is a central dispatching institution with access to all relevant information. This dispatching institution will do the calculation and give instructions to the producers. Therefore, in a vertically integrated utility, OPF can be used for two purposes:1) Dispatching purposes meaning that Equation (7.29) is solved with respect to generation, P_i, in input nodes. 2) Pricing purposes which

Price Calculation Based on Optimal Power Flow (OPF)

means that the price given by Equation (7.29) is used for pricing in consumption nodes.

In order to implement the optimal solution within this traditional framework, prices were used as steering signals towards the consumers while quantities (generating levels) were used towards the generating units, see Figure 7.16. This is of course due to the fact that the utility had no direct control over the quantities consumed by its customers, whereas output from the generating units was controlled. It can also be regarded as a rational solution due to the price-quantity characteristics of the two sides. The production is normally more price sensitive than the consumption, and prices are thus less effective as a control instrument on the generating side.

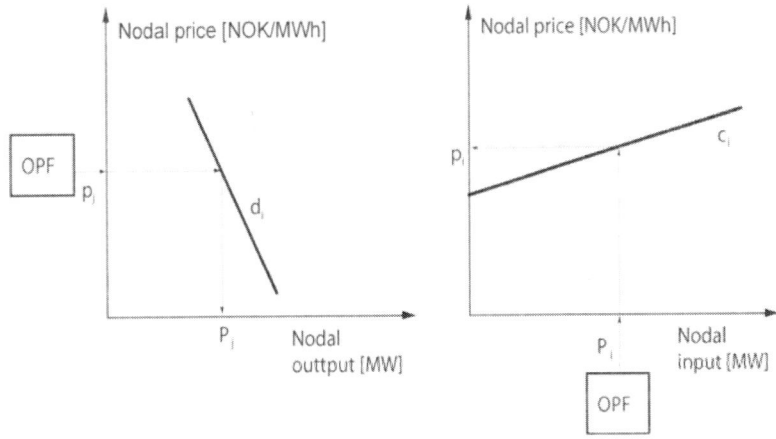

Figure 7.16 OPF generating pricing signals to consumers and quantity signals to generators.

Within the deregulated framework, nodal prices (based on OPF calculation) can be used for input and output and if consumers as well as producers adapt to those prices (as price takers), we will reach the optimal solution.

There are different alternatives for implementation:

1) The nodal price can be applied directly by the responsible institution.[16] That is the single buyer model, see Figure 7.2. This alternative will not involve explicit transmission pricing. The price for transmission will be the difference

---

[16] That institution can be a combination of System Operator, Power pooling institution and grid company.

between the nodal prices of any two nodes. The grid company's income will be the difference between purchase cost at the input nodes and sales income at the output nodes.

2) The nodal prices can be used as the basis for point-to-point transmission prices. The optimal price for transmission from point i to point j is:

$$a_{ij} = p_j - p_i = (\lambda_j - \lambda_i) \tag{7.32}$$

See also Figure 7.3.

3) With a system of point tariffs, each participant is paying for transmission between the connection node and the hub. If one node s, is chosen as the hub, the transmission price between an input bus and this hub will be:

$$a_i = p_s - p_i = p_s - \lambda_i \tag{7.33}$$

and

$$a_j = \lambda_j - p_s \tag{7.34}$$

will be the transmission price between the hub and the output node. The price at the reference node p_s will be the market price.

It is evident that the point tariff at point i plus the point tariff at point j will equal the point-to-point tariff between i and j.

$$a_{ij} = a_i + a_j = (\lambda_j - \lambda_i) \tag{7.35}$$

7.3.3 Marginal transmission losses and nodal prices

There is a link between the transmission tariffs and the grid losses. For the unconstrained case, differences between nodal prices are caused by the grid losses alone. Without losses all the nodal prices would be the same and the transmission prices (based on the principles described here) would be zero.

Price Calculation Based on Optimal Power Flow (OPF)

We consider a grid with total losses T. We change the injection P_i at bus i and assume that is balanced by a corresponding change of input at the hub, s, including loss compensation. All other injections in the system are unchanged.

From the appendix to this chapter we find that:

$$\rho_{ij} = 1 - \frac{\lambda_i}{\lambda_j} \qquad (7.36)$$

where ρ_{ij} is the marginal loss for transfer from node i to node j.

We then define a nodal marginal loss-factor $\rho_i = \partial T / \partial P_i$ for a change in losses in relation to a marginal increase of the transfer between the reference node (the hub) and the node i. Hence:

$$\rho_i = (\lambda_s - \lambda_i)/\lambda_s = 1 - \lambda'_i \qquad (7.37)$$

where we define a relative nodal price: $\lambda'_i = \lambda_i / \lambda_s$.

This means that when no transfer constraints or generation limits are active, the marginal loss-factors can be found once the Lagrange multipliers are known. The ρ_i factors can be found by setting $\lambda_s = 1$ and solving the homogeneous set Equation (7.18) with respect to all the other lambdas, $i = 1, \ldots, n+m$, $i \neq s$.

As mentioned, ρ_i represent the marginal loss factor for node i with respect to the hub, s. If we multiply it with the market price, we get the normal point tariff:

$$a_i = p_s \rho_i = p_s - p_s \lambda'_i = p_s - \lambda_i \qquad (7.38)$$

Notice that we assume a non-congested case here. But the losses are not affected by a congestion if we regard the load flow as given. The nodal prices, however, are affected. Equation (7.18) is therefore only valid for a non-congested case.

7.3.4 Timing and data

Due to the fact that the power system is dynamic and the flow is highly variable, the provision of data for the OPF is a problem. Timing is an important factor. There are two aspects to consider:

Time resolution.

How frequently should prices change? The Norwegian system is based on hourly spot prices, but the transmission energy charges (or more precise: the marginal loss factor) change only twice in a 24 h period. The congestion charge on the other hand, changes hourly in line with the Spot Price

Timing of the calculation and dissemination of prices.

Normally it is required that transmission prices are available some time before real time (ex ante) in such a way that participants are informed and can take it into account in their market behaviour. However, it can be discussed how far ahead calculation and dissemination of prices should be done. In Sweden Finland and Denmark this is done once a year, in Norway every two months (in the future perhaps every week) and in California (before 2001) two days before real time.

As long as the calculations are done ex ante, it has to be based on forecasts and there will be some uncertainty. This uncertainty will increase with increased distance to real time and with increased time resolution.

If the market design is based on central scheduling/dispatch, (see Section 5.7), it is possible to run an OPF as part of the market clearing procedure. In that case the OPF can be based on data provided through the bidding process. That means the time resolution for the transmission charge is the same as that for the market price (the Spot Price). It also means that the calculation is close to real time.

Ex post pricing is also possible. In that case the calculation could be based on observed load flow and could be very precise. But the planning situation for the market participants would be difficult. They would have to make a forecast of transmission prices.

In the Nordic system, the timing is handled differently for the different components in the transmission tariffs. The energy (marginal loss) charge is presented ex ante, while the congestion charge is handled close to real time. Other tariff components are normally adjusted once a year.

There are obvious reasons for this. The congestion charge (or the congestion management) is decisive for system security. If the capacity limit of a transmission line is violated, it can be disastrous. The other tariff components are less critical.

7.3.5 Compensation for transmission losses

It is an implicit assumption here that the grid owner or the System Operator is carrying all the grid costs, including the cost for transmission losses. That means that the grid owner has to buy electricity in the market to compensate for losses, hour by hour. Current practice in Norway and Sweden is slightly different. Statnett

Price Calculation Based on Optimal Power Flow (OPF)

is buying loss energy in the Spot Market while Svenska Kraftnät is making long-term contractual purchases. In both cases they are acting in the market in line with other electricity consumers.

As was pointed out in Section 7.3.3, the optimal transmission price equals the cost of marginal losses (as long as there are no congestions). Because the marginal losses are roughly twice the average losses, the grid company's revenue from the transmission price is about twice the cost for transmission losses (provided the current market price is used for energy losses).

7.3.6 Location of the market place (hub)

A point tariff system is based on the assumption that the producers are paying for transmission from the input nodes to the market place while the consuming side is paying for transfer from the market place to the output node. It is obvious that the market price is influenced by the transmission tariffs. If the tariffs at all input nodes are increased by Δp and there is a corresponding decrease on all the output nodes, the market price will increase by Δp. Hence the local price seen by the market participants will not change.

But this change in market price will have certain consequences for the grid company. The company has to pay market price for the losses, and a high price will increase the cost of that.

In order to analyse the consequences of this type of change, we compare a reference case and a Δp case.

Table 7.3 Definition of a reference case and a Δp case

	Reference case	Δp case
Input from generators	P_i	P_i
Output of consumer	P_j	P_j
Transmission price input	a_i	$a_i + \Delta p$
Transmission price output	a_j	$a_j - \Delta p$
Market price	p_s	$p_s + \Delta p$
Grid losses	T	T
Generation cost	C_i	C_i

Chapter 7: Grid access

The losses are the difference between input and output:

$$T = \sum_{i=1}^{n} P_i - \sum_{i=1}^{m} P_j \qquad (7.39)$$

The consequences for the different parties are shown in Table 7.4.

Table 7.4. Comparison of the reference case and the Δp case

Generating company	Reference case	Δp case
Generator sales income Generator transmission cost Generation cost	$P_i\,p_s$ $P_i\,a_i$ C_i	$P_i\,(p_s+\Delta p)$ $P_i\,(a_i+\Delta p)$ C_i
Generation surplus	$P_i\,(p_s-a_i) - C_i$	$P_i\,(p_s-a_i) - C_i$

Consumer	Reference case	Δp case
Market purchase Transmission cost	$P_j\,p_s$ $P_i\,a_j$	$P_j\,(p_s+\Delta p)$ $P_j\,(a_j-\Delta p)$
Total expenses	$P_j\,(p_s+a_j)$	$P_j\,(p_s+a_j)$

Price Calculation Based on Optimal Power Flow (OPF)

Transmission company	Reference case	Δp case
Purchase of energy (losses)	$T\ p_s$	$T\ (p_s + \Delta p)$
Revenue: From generator	$\Sigma\ P_i\ a_i$	$P_i\ (a_i + \Delta p)$
From consumers	$\Sigma\ P_j\ a_j$	$P_j\ (a_j - \Delta p)$
Increased cost: $T \Delta p$ Increased revenue: $\Sigma\ Pi\ \Delta p - \Sigma\ Pj\ \Delta p = \Delta p\ (\Sigma\ Pi - \Sigma\ Pj) = T \Delta p$		

We can conclude that there will be no economic change for the market participants. The grid company will have the same change of cost and revenue, so the net revenue will be unchanged. So the hub can change location without any economic consequences for the parties.

A few numerical examples are given later in this chapter where we demonstrate this indifference concerning hub location.

7.3.7 Security of supply

Ordinary economic dispatch can be extended to include security of supply. In order to explain basic elements of security-constraint optimal power flow (SECOPF) and possible implications for transmission pricing, the following description illustrated by a simple example is used [4].

We distinguish between four different operating states.

- *Optimal dispatch*: this is the state that the power system is in prior to any contingency. It is optimal with respect to economic operation, but it may not be secure.

- *Post contingency*: is the state of the power system after a contingency has occurred. We assume here that this state has a security violation (line or transformer beyond its flow limit, or a bus voltage outside the limit).

- *Secure dispatch*: is the state of the system with no contingency outages, but with corrections to the operating parameters to account for security violations.

- *Secure post-contingency*: is the state of the system when the contingency is applied to the base-operating condition - with corrections.

We illustrate these concepts with an example. We have a simple power system consisting of two generators, a load and a double circuit line. Both generators are supplying the load as shown below (we ignore losses).

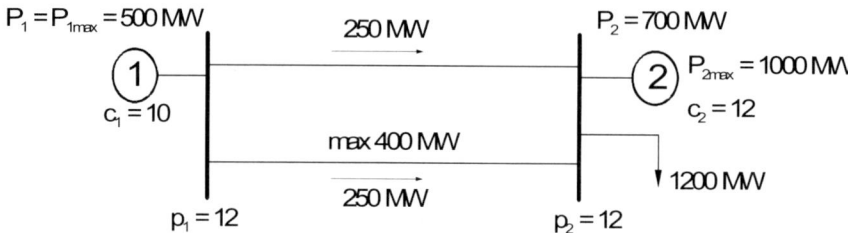

Figure 7.17 Optimal dispatch.

We start with an ordinary economic dispatch implying 500 MW from unit 1 and 700 MW from unit 2. Each circuit of the double circuit line can carry a maximum of 400 MW, thus there is no loading problem in this base-operating condition.

Next we assume an outage of one of the two transmission lines caused by a failure. This results in a post contingency state with overload on the remaining line, shown in Figure 7.18.

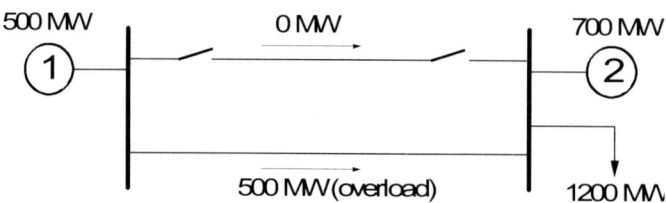

Figure 7.18 Post contingency state.

If we want to avoid this situation, we can reschedule generation. We reduce output from unit 1 to 400 MW and increase output from unit 2 to 800MW. The secure dispatch is as shown in Figure 7.19. Notice that the nodal prices are different from the ordinary economic dispatch.

Price Calculation Based on Optimal Power Flow (OPF)

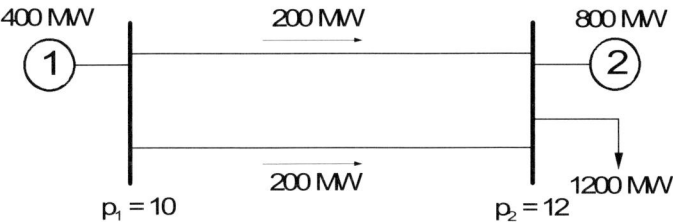

Figure 7.19 Secure dispatch.

The post-contingency condition in this case includes no overload. It is a secure post contingency state as seen in the figure below.

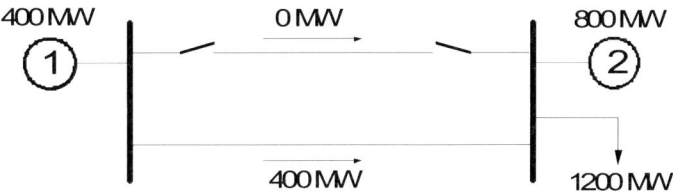

Figure 7.20 Secure post contingency state.

W noticed the difference in nodal prices between the ordinary optimal dispatch and secure dispatch. That can be explained by equation (7.29) (e). Ordinary optimal dispatch involves full output (500 MW) from unit 1. That means the nodal price at that point will be marginal generation cost, c1 plus the marginal value of further capacity, γ_1. At node 2 there will be further available capacity so the nodal price will equal the marginal generation cost c2. That will also be the nodal price at node 1 because transfer from 2 to 1 can be done without losses or restrictions.

$$p_1 = \lambda_1 = c_1 + \gamma_1 = 10 + 2 = 12$$
$$p_2 = \lambda_2 = c_2 = 12$$

In the secure dispatch state, there is idle generating capacity on both sides and the nodal prices must equal the marginal cost:

$$p_1 = \lambda_1 = c_1 = 10$$
$$p_2 = \lambda_2 = c_2 = 12$$

By adjusting the output on unit 1 and unit 2, we have prevented overload in the post-contingency state. In order to achieve this dispatch adjustment, prices have to be adjusted according to this.

The secure dispatch or security-constrained optimal power flows (SECOPF) as it is described here, is based on an (n-1) criterion. That means we can lose one component and full supply can still be maintained. This criterion is common in SECOPF programs [4].

7.3.8 Three-node DC example

In order to demonstrate OPF calculation based on Equations (7.17) – (7.19) (alternatively Equations (7.25) - (7.29) for a congested case), we use a three-node network as an example. To simplify it we use a DC-network.

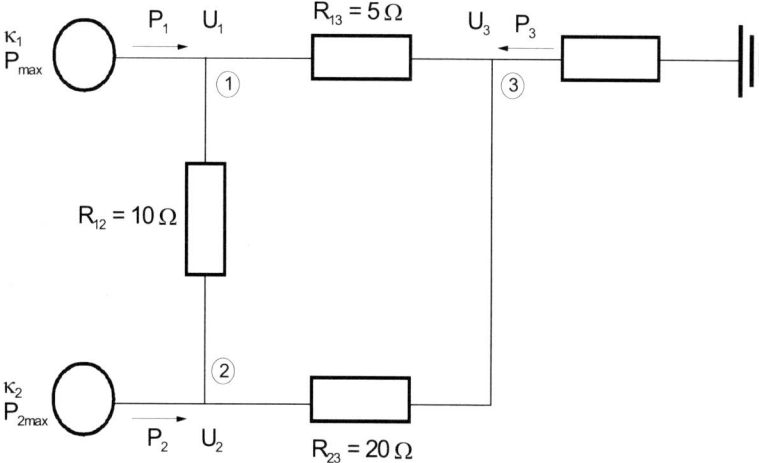

Figure 7.21 Example DC grid.

Price Calculation Based on Optimal Power Flow (OPF)

Generation and consumption data:

Generator 1:   Marginal cost c_1 = 0.12 NOK/kWh

Max generation P_1 max = 20 MW

Generator 2:   Marginal cost c_2 = 0.14 NOK/kWh

Max generation P_2 max = 30 MW

Load: P3 = -30 MW (constant).

Uncongested case

The calculations are shown in the appendix. The results are shown in Figure 7.22.

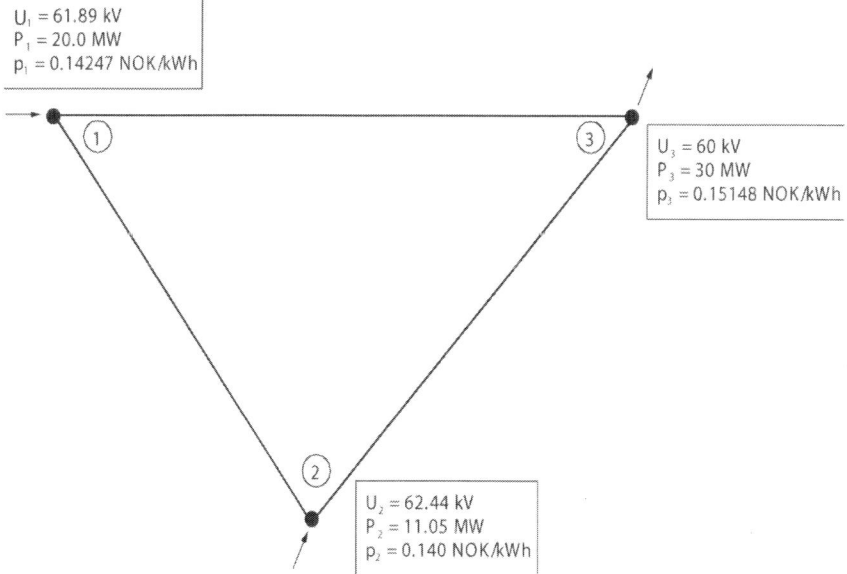

Figure 7.22 Optimal load flow. Uncongested case.

We next regard the calculation of per unit (p.u) marginal loss factors based on alternative hub locations, here in node 3 and node 2. The calculations are showed in the Appendix. Table 7.5 shows the cash flow for the two alternatives.

Table 7.5 Cash flow. Uncongested case.

	Hub at node 3	Hub at node 2
Spot price p_s (NOK/kWh)	0.15148	0.140
Nodal "loss" factors:		
ρ_{1in}	0.059	-0.018
ρ_{2in}	0.076	0.0
ρ_{3out}	0.0	0.082
Generator 1		
Sales $P_1 p_s$ (NOK/h)	3030	2800
- transmission $P_1 \rho_{1in} p_s$ (NOK/h)	- 179	+ 50
Net income (NOK/h)	2851	2850
Generator 2		
Sales $P_2 p_s$ (NOK/h)	1674	1547
- transmission $P_2 \rho_{2in} p_s$ (NOK/h)	- 127	0
Net income	1547	1547
Load		
Purchase $P_3 p_s$ (NOK/h)	4544	4200
- transmission $P_3 \rho_{3out} p_s$ (NOK/h)	0	344
Total	4544	4544
Grid company		
Transmission gen 1 (NOK/h)	179	-50
Transmission gen 2 (NOK/h)	127	0
Transmission load (NOK/h)	0	344
- losses $P_{loss} \cdot p_s$ (NOK/h)	- 159	- 147
Net revenue (NOK/h)	147	147

These examples illustrate that alternative locations of the hub have an influence on the market price and on the transfer price, but no impact on the net economic outcome for the entities involved.

Congested three-node example

We now extend the example by introducing a congestion between nodes 1 and 3. Maximum current is set to 350 A which means that

Price Calculation Based on Optimal Power Flow (OPF)

$$I_{1,3} = \frac{1}{R_{31}}(U_1 - U_3) \leq I_{1,3\max} \quad (7.42)$$

or

$$U_1 - U_3 \leq R_{31} \cdot I_{3\max} = 1.75 \quad (7.43)$$

The calculation is shown in the appendix and the solution is shown in Figure 7.23 Figure 7.23.

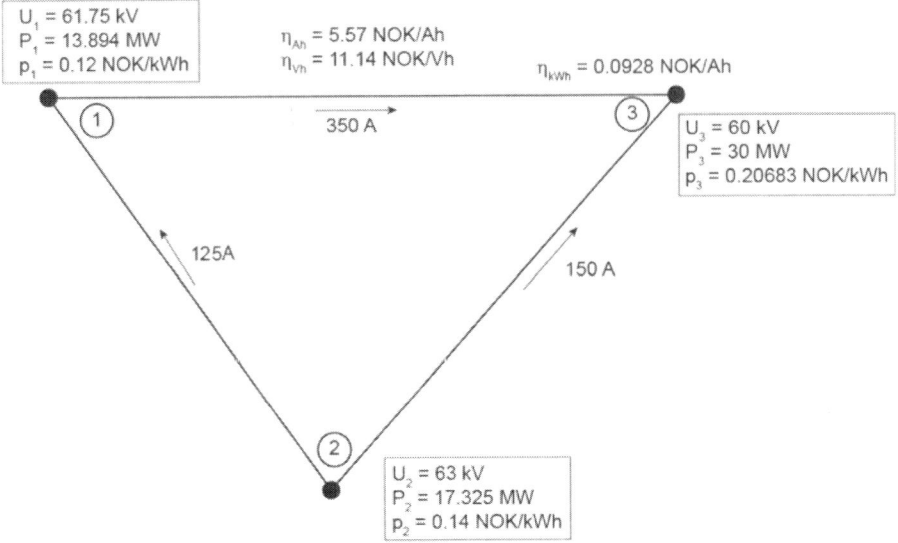

Figure 7.23 Optimal load flow. Congested case.

If we compare this solution with the uncongested case, we see that there has been a shift of generation from the cheap generator 1 to the more expensive generator 2. The higher price in node 3 is caused by the fact that an increased load in this point would lead to a further shift from the cheap to the more expensive generator.

It is worth noticing that in this congested case, there is no simple link between nodal prices and marginal grid losses.

One interesting observation is that the energy flow between node 2 and 1 is from a high to a low price node. From an economic point of view, this seems irrational.

But it is in line with the economic optimum solution. (Similar examples are described in e.g. [3] and [5]).

7.3.9 Three-node DC Equivalent

The example is taken from [5]. The problem is described in the two figures below. We have two nodes with production and one node with consumption. We assume a reference voltage of 100 kV, lines without losses, linearly increasing marginal costs in the feed-in nodes 1 and 2 and a constant willingness to pay in node 3.

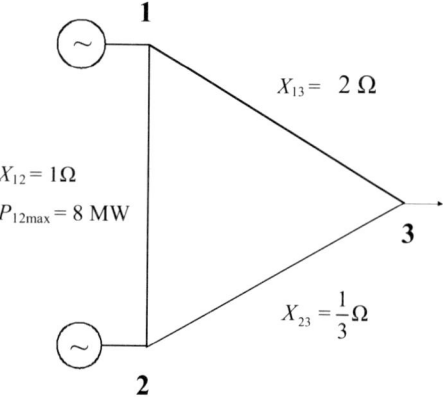

Figure 7.24 A three-node DC system with two production nodes and one consumption node.

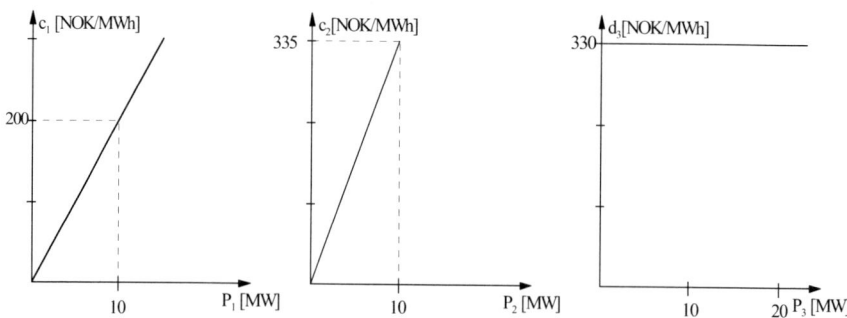

Figure 7.25 Marginal production costs in nodes 1 and 2 and marginal willingness to pay in node 3.

Price Calculation Based on Optimal Power Flow (OPF)

Unrestricted

As a first approach we neglect all transmission constraints. Due to the fact that there are no losses in the lines the nodal prices are the same in all nodes and equal to the willingness to pay in node 3, i.e. 330 NOK/MWh. This is the *System Marginal Cost* (SMC), sometimes called *System Lambda*.

Under these assumptions the production costs in node 1 and 2 determine the allocation of the production, P_1=16.5 MW and P_2=9.85 MW. Consumption in node 3 is simply the sum of the two, $P_3=P_1+P_2$=26.35 MW.

To find the transmitted energy in the different lines we need the transmission equations. The energy transmitted between two nodes when we have no losses is

$$P_{ij} = \frac{U_i}{X_{ij}}(\delta_i - \delta_j) \tag{7.44}$$

where U_i is the voltage, X_{ij} is the reactance in the line and δ_i is the voltage angle.

The power fed into a node can be expressed as the sum of the transmitted power in the lines out of the node, and this is equivalent to equation:

$$g(x, P) = 0$$

where

$$x = \begin{bmatrix} \delta_1 \\ \delta_2 \\ \delta_3 \end{bmatrix}$$

which can be written:

$$\frac{U_1}{X_{12}}(\delta_1 - \delta_2) + \frac{U_1}{X_{13}}(\delta_1 - \delta_3) - P_1 = 0$$

$$\frac{U_2}{X_{21}}(\delta_2 - \delta_1) + \frac{U_2}{X_{23}}(\delta_2 - \delta_3) - P_2 = 0 \qquad (7.45)$$

$$\frac{U_3}{X_{31}}(\delta_3 - \delta_1) + \frac{U_3}{X_{32}}(\delta_3 - \delta_2) - P_3 = 0$$

The voltages (voltage levels) are equal (100 kV which is defined as 1 p.u.) and we set the voltage angle in node 1 to zero:

$$\delta_1 = 0$$

This gives three equations and two unknowns, and we find that

$$\delta_2 = -0.08915 \text{ rad and } \delta_3 = -0.1517 \text{ rad}$$

The transmitted power in the lines can be found from Equation(0.1). The numerical values are

$$P_{12} = 8.915 \text{ MW}, \quad P_{13} = 7.585 \text{ MW and } P_{23} = 18.756 \text{ MW}$$

The socio-economic surplus is given by the consumer surplus plus the producer surplus. In our example the consumer surplus equals zero. The producer surplus equals the area between the price and the marginal cost:

$$CS_1 = \frac{1}{2} \cdot 16.5 \cdot 330 = 2722.50 \text{ NOK/h}$$

$$CS_2 = \frac{1}{2} \cdot 9.85 \cdot 330 = 1625.25 \text{ NOK/h}$$

The socio-economic surplus then becomes:

$$SES = CS_1 + CS_2 = 4347.25 \text{ NOK/h}$$

Price Calculation Based on Optimal Power Flow (OPF)

Transfer Restriction

We introduce a transfer restriction on the line between nodes 1 and 2 on 8 MW. The objective is to maximize the social economic surplus within the technical constraints.

$$\text{Max}[SES]$$

s.t.

$$P_1 - \frac{U_1}{X_{12}}(\delta_1 - \delta_2) - \frac{U_1}{X_{13}}(\delta_1 - \delta_3) = 0$$

$$P_2 - \frac{U_2}{X_{21}}(\delta_2 - \delta_1) - \frac{U_2}{X_{23}}(\delta_2 - \delta_3) = 0$$

$$-P_3 - \frac{U_3}{X_{31}}(\delta_3 - \delta_1) - \frac{U_3}{X_{32}}(\delta_3 - \delta_2) = 0$$

$$P_{12} \leq P_{12\max} \Rightarrow P_{12\max} - \frac{U_1}{X_{12}}(\delta_1 - \delta_2) = 0$$

(7.47)

The belonging Lagrange function from the Karush-Kuhn-Tucker is

$$L = CU - (C_1 + C_2) + \lambda_1[P_1 - \frac{U_1}{X_{12}}(\delta_1 - \delta_2) - \frac{U_1}{X_{13}}(\delta_1 - \delta_3)]$$

$$+ \lambda_2[P_2 - \frac{U_2}{X_{21}}(\delta_2 - \delta_1) - \frac{U_2}{X_{23}}(\delta_2 - \delta_3)]$$

$$+ \lambda_3[-P_3 - \frac{U_3}{X_{31}}(\delta_3 - \delta_1) - \frac{U_3}{X_{32}}(\delta_3 - \delta_2)]$$

$$+ \eta_{12}[P_{12\max} - \frac{U_1}{X_{12}}(\delta_1 - \delta_2)]$$

(7.48)

where λ_i is the nodal prices and η is the shadow price for the transfer constraint. By setting the derivatives with respect to P_i, λ_i, δ_i and η equal to zero we get an equation system that can be solved. The solution is given in Figure 7.26. We see that the power flows from node 1 to node 2 – from a low price node to a high price node. Again this might seem to be in violation of basic economic laws. However, the solution is the one that both maximizes the socio-economic surplus and satisfies Kirchhoff's laws.

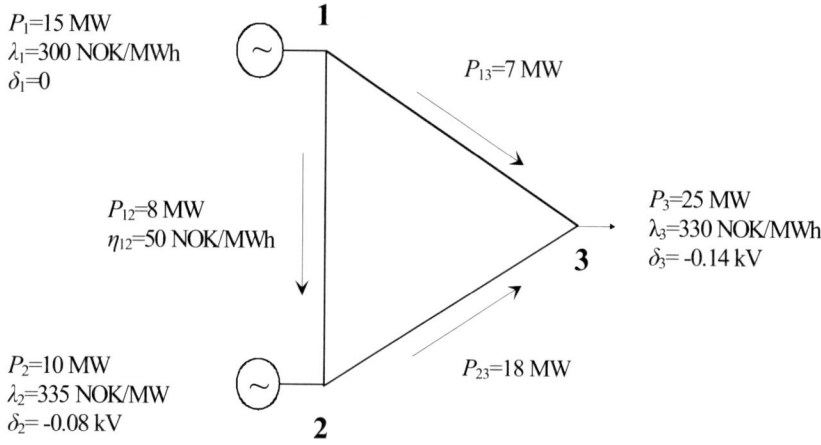

Figure 7.26 The numerical solution to the problem.

By assuming a single buyer model (Section 7.1), we can find the grid owner's income:

$$R = P_3\lambda_3 - P_1\lambda_1 - P_2\lambda_2 \qquad (7.49)$$

With numerical values this gives $R = 400$ NOK/h. Alternatively one can chose a reference price and calculate an input and an output tariff as the difference between the reference price and the nodal prices. A third alternative is to multiply the power flow on the lines with the price difference between the corresponding nodes.

The surplus to the producers is the revenue minus the cost. The revenue is the price multiplied by produced power and the cost is given by the area under the marginal cost curve. The producer surplus for the two producers are:

$$PS_1 = \frac{1}{2}\lambda_1 P_1 = \frac{1}{2}300 \text{NOK/MWh} \cdot 15\text{MW} = 2250\text{NOK/h}$$

$$PS_2 = \frac{1}{2}\lambda_2 P_2 = \frac{1}{2}335 \text{NOK/MWh} \cdot 10\text{MW} = 1675\text{NOK/h}$$

Price Calculation Based on Optimal Power Flow (OPF)

The consumer surplus is zero because all the consumers pay exactly what they are willing to pay. Hence the total economic surplus is the sum of the producer surpluses plus the revenue to the grid owner (the congestion rent):

$$PS_1 + PS_2 + R = 2250 \text{NOK/h} + 1675 \text{NOK/h} + 400 \text{NOK/h} = 4325 \text{NOK/h}$$

We notice a small loss compared to the uncongested case.

7.3.10 Hydro and Hydro-thermal Systems

OPF is a static optimization technique. It can be applied to scheduling problems for thermal systems without too much error. However, it is not applicable to hydropower scheduling. In this section a simple example is used to discuss transmission pricing in a hydropower context. Despite its simplicity it should be possible to draw some general conclusions from the examples.

Only hydropower

Two power plants A and B are connected to a load L over two individual transmission lines (see Figure 7.27 and Figure 7.28). The transmission losses are proportional to R_A and R_B and increase with the square of transferred power. Losses are higher from A than from B: $R_A > R_B$. The demand is modelled as price dependent and variable over time with a typical day and night demand as shown in Figure 7.28. To make it simple we assume the day and night duration to be 12 h each.

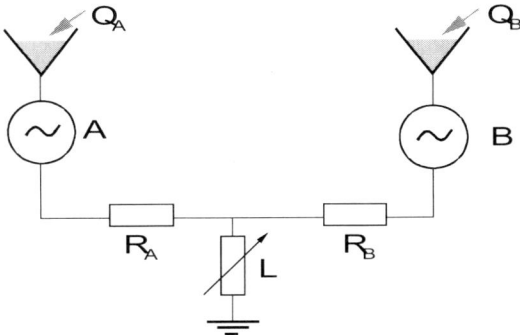

Figure 7.27 Two hydropower plants covering one load.

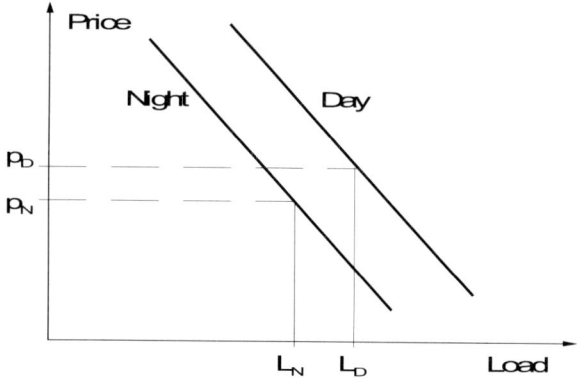

Figure 7.28 Day and night demand.

The following variables are defined:

L_D - day time load [kWh]

L_N - night time load [kWh]

p_D - day price (referred to the load point) [NOK/kWh]

p_N - night price (referred to the load point) [NOK/kWh]

R_A - specific loss line A [(kWh)$^{-1}$]

R_B - specific loss line B [(kWh)$^{-1}$]

L_{AD} - delivery from A daytime [kWh]

L_{AN} - delivery from A nighttime [kWh]

L_{BD} - delivery from B daytime [kWh]

L_{BN} - delivery from B nighttime [kWh]

Q_A - inflow to reservoir A [kWh/24 h]

Q_B - inflow to reservoir B [kWh/24 h]

The demand is modelled as a linear function of price,

$$p_D = K_D - k L_D \quad (0.2)$$
$$p_N = K_N - k L_N \quad (0.3)$$

where K_D, K_N and k are constants.

Price Calculation Based on Optimal Power Flow (OPF)

We assume that day plus night time generation equals inflow, which means that long run reservoir contents remain unchanged. We further assume that there are no active restrictions on reservoir capacity. This leads to the following energy balance equations (we assume 100% generating efficiency),

$$L_{AN} + R_A L_{AN}^2 + L_{AD} + R_A L_{AD}^2 = Q_A \quad (7.52)$$
$$L_{BN} + R_B L_{BN}^2 + L_{BD} + R_B L_{BD}^2 = Q_B \quad (7.53)$$

that are the restrictions to our optimization problem.

The objective function we wish to maximize is based on the same principles as Equation (7.7). The difference is that there are no costs involved because it is assumed that variable operation costs are zero and that there are no costs involved in moving generation between day and night. The objective function is given by the area under the demand curve (i.e. consumer surplus).

$$\max \left[\int_0^{L_D} p_D dL + \int_0^{L_N} p_N dL \right] \quad (7.54)$$

The optimization problem can be solved by introducing Lagrange multipliers and extend the objective function:

$$H = \int_0^{L_D} p_D dL + \int_0^{L_N} p_N dL \quad (7.55)$$

$$- \lambda_A [L_{AN} + R_A L_{AN}^2 + L_{AD} + R_A L_{AD}^2 - Q_A]$$
$$- \lambda_B [L_{BN} + R_B L_{BN}^2 + L_{BD} + R_B L_{BD}^2 - Q_B]$$

where

$$L_{AN} + L_{BN} = L_N \quad (7.56)$$
$$L_{AD} + L_{BD} = L_D \quad (7.57)$$

First order conditions for optimum are found in the usual way by setting the partial derivatives with respect to the variables, L_{AN}, L_{AD}, L_{BN}, L_{BD}, λ_A and λ_B, to zero.

$$\frac{\partial H}{\partial L_{AN}} = 0 \Rightarrow p_N - \lambda_A (1 + 2 R_A L_{AN}) = 0 \quad (7.58)$$

$$\frac{\partial H}{\partial L_{BN}} = 0 \Rightarrow p_N - \lambda_B(1 + 2R_R L_{BN}) = 0 \qquad (7.59)$$

$$\frac{\partial H}{\partial L_{AD}} = 0 \Rightarrow p_D - \lambda_A(1 + 2R_A L_{AD}) = 0 \qquad (7.60)$$

$$\frac{\partial H}{\partial L_{BD}} = 0 \Rightarrow p_D - \lambda_B(1 + 2R_B L_{BD}) = 0 \qquad (7.61)$$

$$\frac{\partial H}{\partial \lambda_A} = 0 \Rightarrow eq.(7.52)$$

$$\frac{\partial H}{\partial \lambda_B} = 0 \Rightarrow eq.(7.53)$$

This set of equations is then solved with respect to the six unknowns. Because of the square elements a computer program is used to find a numerical solution, based on the following input data.

k	= 0.001	[NOK/kWh/MW]
K_N	= 0.19	[NOK/kWh]
K_D	= 0.22	[NOK/kWh]
R_A	= 0,002	[1/kWh]
R_B	= 0,0008	[1/kWh]
Q_A	= Q_B = 100 MWh	

Price Calculation Based on Optimal Power Flow (OPF)

Table 7.6. Optimal schedule and corresponding prices.

	Day	Night
Generation plant A (MWh)	54.86	45.14
Generation plant B (MWh)	60.07	39.93
Price plant A (NOK/kWh)	0.0939	0.0939
Price plant B (NOK/kWh)	0.1032	0.1032
Load (MWh)	107.3	80,4
Price at load point (NOK/kWh)	0.1127	0.1096
Losses (MWh)	7.63	4.67

We notice the following:

- There is no price difference between day and night at the power plants. The prices can be interpreted as the local shadow price of water (water values). As long as there is no cost involved in transfer of generation between day and night, and no active constraints on the reservoirs, there will not be any price difference between day and night.

- The difference between day and night generation is larger for plant B than for plant A. This is caused by the larger loss on transfer from A. The plant with lowest transmission losses carries most of the day/night generation variability.

- The price difference between the load node and the generating nodes equals the marginal losses.

As in the previous section we have to select a market place or in fact a market price in order to calculate point tariffs based on this optimal schedule. We investigate two alternatives in order to illustrate the point that this choice neither has economic impact on the grid customers, nor on the grid operator.

The producers are assumed to be profit-maximizing entities. They have a limited amount of water available and in this case their only freedom of choice lies in the allocation of day and night generation. If there is a local price difference between day and night, they will produce more in the high price period and vice versa. The optimal schedule (see Table 7.6) is characterized by price equality between the two periods and the producers have no motive to depart from the optimal schedule. If the prices changes by an equal amount in both periods, the producers will continue to use the same schedule. This means that we can add a constant (positive or negative) to the point tariff paid by a producer without affecting his generation (as long as the local price is above zero).

Chapter 7: Grid access

Hydro-thermal examples

Three different hydro-thermal examples are investigated. We use the same demand and grid loss characteristics as in the previous hydro example. The only difference is that one of the hydro plants is replaced by a thermal one. The cases differ in the cost of the thermal generation, which is indicated in the upper right-hand box in Figure 7.29 - Figure 7.31. In the same figures the optimum schedule is shown with output and nodal prices for the two plants and nodal prices at the load node.

Note the following:

- For the hydro plant the prices for day and night are always equal (as long as there are no binding constraints on generation or reservoir capacity).

- For the thermal plant, the price equals the marginal cost unless the output is affected by a capacity constraint.

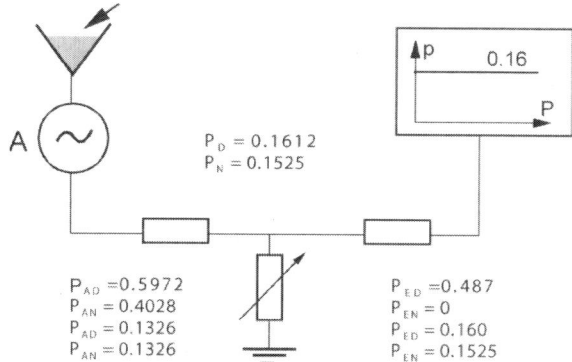

Figure 7.29 *Hydro and thermal plant (with constant marginal cost) covering one load*

Price Calculation Based on Optimal Power Flow (OPF)

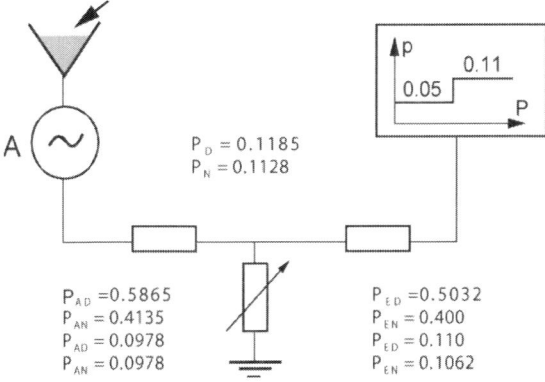

Figure 7.30 Hydro and thermal power plant, with variable marginal cost, covering one load.

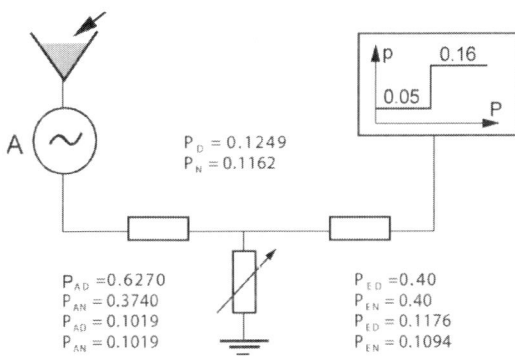

Figure 7.31 Hydro and thermal plant, with variable marginal cost, covering one load.

7.3.11 DC–equivalent. 11 node (zone) power system

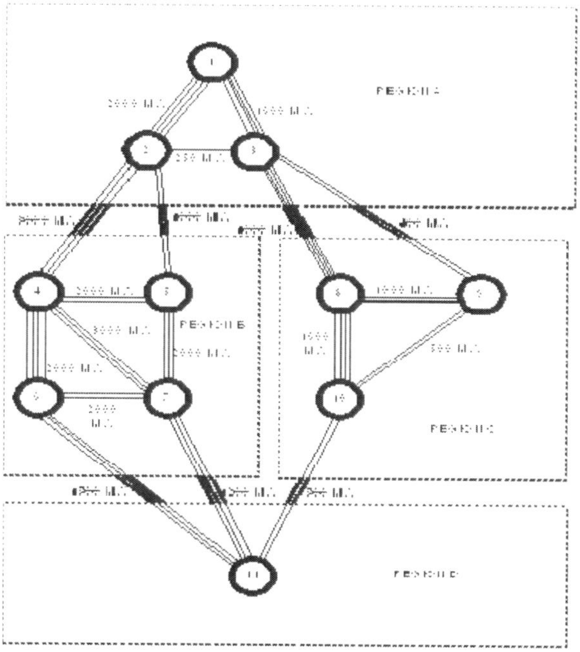

Figure 7.32 Eleven node grid.

The example described in this section is taken from [1]. A full description of data, procedure and results is given in that paper.

The example is an eleven-zone power system, shown in Figure 7.32. Each zone is a collection of electrical buses connected with enough capacity so that overloads on transmission lines and transformers within the zone can be neglected. Thus, each zone can be treated as a single bus in power flow computation. Zones are connected by interfaces each consisting of multiple identical transmission lines (circuits). Power flow limits (ratings) in MW for the interfaces are given in

Table 7.7. The rating of each interface is simply the rating of each circuit in the interface times the number of circuits in the interface. The line reactances in the example power system are given in

Table 7.7.

Table 7.7 Example transmission system data.

Interface	From Zone	To Zone	No. of Circuits	Circuit Reactance X, per unit	Capacity in MW
1	1	2	4	0.020	2000.0
2	1	3	4	0.025	1600.0
3	2	3	2	0.080	250.0
4	2	4	3	0.010	3000.0
5	2	5	2	0.020	1000.0
6	3	8	4	0.040	1000.0
7	3	9	2	0.050	400.0
8	4	5	2	0.010	2000.0
9	4	6	4	0.020	2000.0
10	4	7	3	0.010	3000.0
11	5	7	3	0.015	2000.0
12	6	7	2	0.010	2000.0
13	8	10	4	0.025	1600.0
14	8	9	3	0.030	1000.0
15	9	10	2	0.040	500.0
16	6	11	3	0.020	1500.0
17	7	11	3	0.025	1200.0
18	10	11	2	0.040	500.0

There are generators and loads connected to each node. The generating cost functions are quadratic, which means a linear increase of marginal costs with output. Loads are price inelastic for the cases presented here.

Generating cost functions:

$$C_i(P_{G_i}) = \alpha_i + \beta_i P_{G_i} + \gamma_i P_{G_i}^2 \qquad (7.62)$$

Where α, β and γ are constants that are listed in Table 7.8. The constant α does not affect the solution and is set to zero. The marginal costs can be expressed as:

$$\frac{dC_i}{dP_{G_i}} = \beta_i + 2\gamma_i P_{G_i} \qquad (7.63)$$

The demand in each node has zero price elasticity and is equal to 1000 MW in each node except node 11 where it is 1500 MW.

Table 7.8 Data for generating units

Unit Number	Zone	β Constant	γ Constant	Max MW
1	1	10.00	0.0040	1000.0
2	2	15.00	0.0060	800.0
3	3	50.00	0.0080	1500.0
4	4	12.00	0.0050	2500.0
5	5	15.50	0.0060	1500.0
6	6	15.50	0.0070	1500.0
7	7	21.50	0.0080	1500.0
8	8	16.00	0.0060	1500.0
9	9	14.00	0.0050	1500.0
10	10	13.00	0.0040	1500.0
11	11	16.00	0.0060	700.0
12	11	31.00	0.0090	2000.0

DC Power Flow

Calculating the power flows in a power system from a given set of loads and generator power outputs is an analytical technique that is central to transmission management. A full AC power flow is the most accurate calculation, but its complexity can obscure the intuitive understanding of the relationship between the variables.

The DC power flow model assumes that only the angles of the complex bus voltages vary, and that the variation is small. Voltage magnitudes are assumed to be constant. Transmission lines are assumed to have no resistance, and therefore no losses. These assumptions create a model that is a reasonable first approximation for the real power system, which is only slightly non-linear in normal steady state operation. The model has advantages due to the speed of computation and also has some other useful properties:

Price Calculation Based on Optimal Power Flow (OPF)

- Linearity: If the power in a transaction from one zone to another is doubled, the flows that are directly attributable to this transaction will also double.

- Superposition: The flows on the interfaces can be broken down into a sum of components that are each directly attributable to a transaction in the system.

Examples of OPF solutions

Base Case (no congestion)

Table 7.9 and Table 7.10 give the base case OPF solution. The entire load is covered and all the generators are producing except the generator in zone 3 and the second generator in zone 11 which are so expensive that they are not used at all. Note that any generator that is not at its minimum or maximum operates at the same incremental cost.

Table 7.9 OPF solution. Base case.

Zone Number	Generation (MW)	Load (MW)	Zone Lambda (USD/MWh)	Generator Marginal Cost (USD/MWh)
1	1000.0	1000.0	30.64	18.00
2	800.0	1000.0	30.64	24.60
3	0.0	1000.0	30.64	50.00
4	1864.4	1000.0	30.64	30.64
5	1262.0	1000.0	30.64	30.64
6	1081.7	1000.0	30.64	30.64
7	571.5	1000.0	30.64	30.64
8	1220.3	1000.0	30.64	30.64
9	1500.0	1000.0	30.64	29.00
10	1500.0	1000.0	30.64	25.00
11	700.0	1500.0	30.64	24.40
Total	11500.0	11500.0		

In the base case all zones have the same zone price (λ). Note that zone 11 is importing 800 MW of power, its first generator is at its maximum output of 700 MW and its second generator is not producing anything. This is an example of a system with no congestion. The flows on this system are given in Table 7.10.

Table 7.10 Base case transmission system flows.

Path	From	To	Low (MW)	Flow (MW)	High (MW)	Percent Loading
1	1	2	-2000.0	-111.7	2000.0	5.6
2	1	3	-1600.0	111.7	1600.0	7.0
3	2	3	-250.0	31.4	250.0	12.6
4	2	4	-3000.0	-267.6	3000.0	8.9
5	2	5	-1000.0	-75.5	1000.0	7.5
6	3	8	-1000.0	-562.8	1000.0	56.3
7	3	9	-400.0	-294.1	400.0	73.5
8	4	5	-2000.0	27.5	2000.0	1.4
9	4	6	-2000.0	206.9	2000.0	10.3
10	4	7	-3000.0	362.3	3000.0	12.1
11	5	7	-2000.0	214.1	2000.0	10.7
12	6	7	-2000.0	34.6	2000.0	1.7
13	8	10	-1600.0	-169.8	1600.0	10.6
14	8	9	-1000.0	-172.6	1000.0	17.3
15	9	10	-500.0	33.2	500.0	6.6
16	6	11	-1500.0	254.1	1500.0	16.9
17	7	11	-1200.0	182.5	1200.0	15.2
18	10	11	-500.0	363.4	500.0	72.7

Figure 7.33 provides an illustration of the solution for the first four nodes. It shows the marginal costs for the generators as a function of net power flow. The generating units in nodes 1 and 2 produce at full capacity because the marginal cost is lower than the nodal price. Unit 3 has zero production because the marginal cost is higher than nodal price. For unit 4 the production level is decided by the point where the marginal cost equals the nodal price.

Price Calculation Based on Optimal Power Flow (OPF)

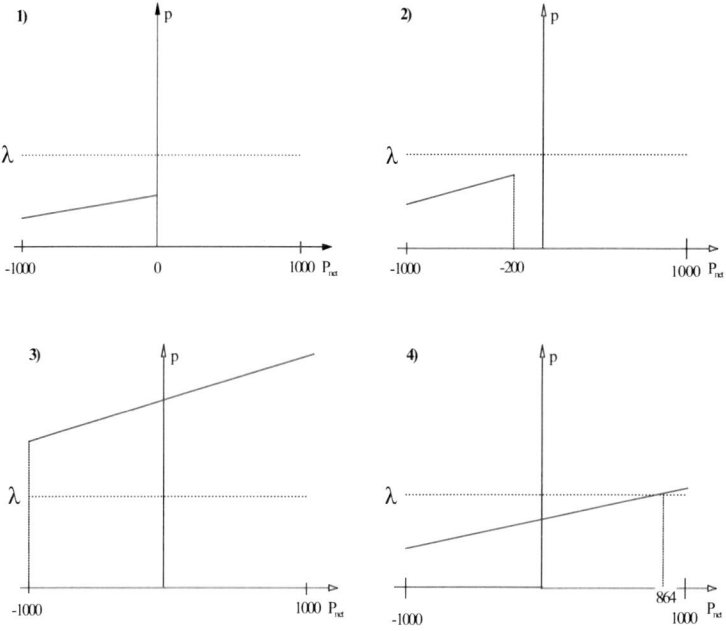

Figure 7.33 Marginal costs as a function of net power flow for nodes 1 to 4.

Congested case. Radial link.

Without any changes in the input/output characteristics of the base case, congestion is created by a change in the transmission system topology. All circuits in the interfaces between zones 6 and 11 and zones 7 and 11 are out of operation. The resulting congested system is shown in Table 7.11.

Table 7.11 Congestion on radial link (from node 10 to 11).

Zone Number	Generation (MW)	Load (MW)	Zone Lambda (USD/MWh)
1	1000.0	1000.0	29.33
2	800.0	1000.0	29.33
3	-0.0	1000.0	29.33
4	1733.4	1000.0	29.33
5	1152.8	1000.0	29.33
6	988.1	1000.0	29.33
7	489.6	1000.0	29.33
8	1111.1	1000.0	29.33
9	1500.0	1000.0	29.33
10	1500.0	1000.0	29.33
11	1225.0	1500.0	40.45
Totals	11500.0	11500.0	

The active constraint is a contingency of one circuit in the zone 10 to zone 11 interface, which brings the remaining circuit to its flow limit. The congestion results in a reduction of import into zone 11 from 800 MW to 275 MW. This means that generation in zone 11 must increase from 700 MW to 1225 MW in order to supply the load of zone 11, and this increased production has to come from the very high priced second generator in zone 11. The reduction of 325 MW in generation exported from the other zones slightly lowers their zone lambdas to USD 29.33/MWh while zone 11 experiences an increase to USD 40.45/MWh. Notice that all zone lambdas are still equal except for zone 11. That is due to the fact that the congested interface is a radial one. A congestion in a meshed network has a different impact on the zone lambdas.

Congestion in a mashed network

The changes to the network that caused congestion in the cases described above, made the interface from zone 10 to zone 11 radial. (If that interface was removed, the system would be split into two disconnected islands.) When congestion occurred on the radial interface, only two distinct zone prices appeared, one on each side of the interface.

When congestion appears on an interface which is part of a networked (meshed or looped) system, all the zone prices are unique. Congestion in any interface in a networked system affects zone prices in the entire networked system. This is illustrated by restoring the interface from zone 7 to zone 11 and letting the

Price Calculation Based on Optimal Power Flow (OPF)

interface from zone 6 to zone 11 be the only one out of service. The OPF results are given in Table 7.12.

Table 7.12 Congestion in a meshed network.

Zone Number	Generation (MW)	Load (MW)	Zone Lambda (USD/MWh)
1	1000.0	1000.0	30.44
2	800.0	1000.0	30.65
3	-0.0	1000.0	30.17
4	1878.1	1000.0	30.78
5	1274.4	1000.0	30.79
6	1094.9	1000.0	30.83
7	585.9	1000.0	30.87
8	1148.6	1000.0	29.78
9	1500.0	1000.0	29.80
10	1500.0	1000.0	29.53
11	718.1	1500.0	31.33
Total	11500.0	11500.0	

Because of the increased interface capacity to zone 11, more power is imported and the more expensive generator in zone 11 now operates at only 18.1 MW. This lowers the zone 11 price below the level at which the price-elastic load would reduce purchases and thus the load in zone 11 is 1500 MW.

The capacity on the interface from zone 10 to zone 11 is still the active constraint, but this interface is now part of a networked system. As a result, every zone price is unique. According to Kirchhoff's Voltage Law, a change in load or generation in any zone affects the flow at the congested interface, even though the changed load or generation is in a zone far away from the interface. Higher zone prices appear where an increase in output increases the flow on the congested interface and vice versa. Thus, some zones have higher prices than the radial congestion case, and some have lower prices.

Chapter 7: Grid access

11 node AC grid

The same 11 node (zone) grid is used in this example, but the OPF calculation is extended to a full AC calculation[17]. Grid data are shown in Table 7.13. Load and generation data are the same as in the previous case.

Table 7.13 Example transmission system data.

Interface	From Zone	To Zone	No. of Circuits	Circuit Reactance (per unit)		Capacity in MW
				R	X	
1	1	2	4	0.5	0.8	2000.0
2	1	3	4	1.0	1.2	1600.0
3	2	3	2	2.0	3.2	250.0
4	2	4	3	0.3	0.4	3000.0
5	2	5	2	0.5	0.8	1000.0
6	3	8	4	1.0	1.6	1000.0
7	3	9	2	1.5	2.0	400.0
8	4	5	2	0.3	0.4	2000.0
9	4	6	4	1.0	0.8	2000.0
10	4	7	3	0.3	0.4	3000.0
11	5	7	3	0.5	0.8	2000.0
12	6	7	2	0.75	0.8	2000.0
13	8	10	4	1.0	1.2	1600.0
14	8	9	3	1.0	1.2	1000.0
15	9	10	2	1.0	1.6	500.0
16	6	11	3	0.75	0.8	1500.0
17	7	11	3	1.0	1.2	1200.0
18	10	11	2	1.0	1.6	500.0

---

[17] The calculation is based on a program developed by Arne Johannesen, SINTEF Energy Research.

Table 7.14 Results from AC OPF. Base case.

Zone Number	Generation (MW)	Load (MW)	Zone Lambda (USD/MWh)	Marginal cost (USD/MWh)
1	1000.0	1000.0	30.70	18.00
2	800.0	1000.0	30.68	24.60
3	0.0	1000.0	30.75	50.00
4	1864.8	1000.0	30.65	30.65
5	1262.9	1000.0	30.66	30.66
6	1085.0	1000.0	30.69	30.69
7	574.3	1000.0	30.69	30.69
8	1217.4	1000.0	30.61	30.61
9	1500.0	1000.0	30.57	29.00
10	1500.0	1000.0	30.58	25.00
11	700.0	1500.0	30.74	24.40
Total	11504.5	11500.0		

We notice some striking differences compared with the DC equivalent.

Total generation is slightly higher than total consumption. This is due to grid losses.

Zone lambdas are different, even in this non-congested case, again due to losses between nodes.

7.4 Transmission/Distribution (T/D) tariffs

7.4.1 General requirements

Previous sections have discussed transmission pricing in one single grid on the basis of economic efficiency. This section includes both the transmission and distribution grid and we assume different owners are involved. In addition we discuss tariffs, not only prices.

The following requirements are normally put on the (T/D) tariffs:

- The tariffs should facilitate market transactions without practical difficulties
- The tariffs should secure economic efficiency, i.e. give correct pricing signals to the grid users
- The tariffs should provide sufficient income to the grid owner.

- The tariffs should give a fair distribution of costs among grid customers.

In addition, it is sometimes claimed that the tariff should give correct incentives to the grid owner concerning the operation and investment in the grid. However, this is a complex issue and will not be discussed in any detail here. It is assumed previously in this chapter, that the grid company has an objective related to socio-economic efficiency. We have to take into account, however, that the tariff must provide sufficient income to cover the cost of the grid, including some return on investments. With respect to this, the regulatory arrangement is important and this matter will be discussed in chapter 9.

The main focus in this section will be on practical implementation of T/D tariffs in the Nordic grid.

7.4.2 The Nordic T/D tariff system

General aspects

In Norway, Sweden and Finland the main transmission grid is divided from generation and full Third Party Access (TPA) is introduced on all grid levels. Transmission/distribution pricing is based on a point-of-connection tariff system.

The basic idea of a point tariff is that the grid user is charged only according to the node where it is connected. The charge depends on the grid user's output from or input to the grid.

It follows from the point tariff principle that a grid user only relates to one grid owner. That is important in a decentralized system with a multi-level ownership structure.

The Norwegian grid, which is typical for this structure, has three levels:

- A central grid (132 - 400 kV)
- Regional grids (60 - 132 kV)
- Local distribution grids (< 22 kV)

Statnett is responsible for the central grid. Regional utilities are normally responsible for the regional grids and local distribution utilities for the distribution grids (see Chapter 5). In some cases the local and regional grids are integrated in one utility.

Transmission/Distribution (T/D) tariffs

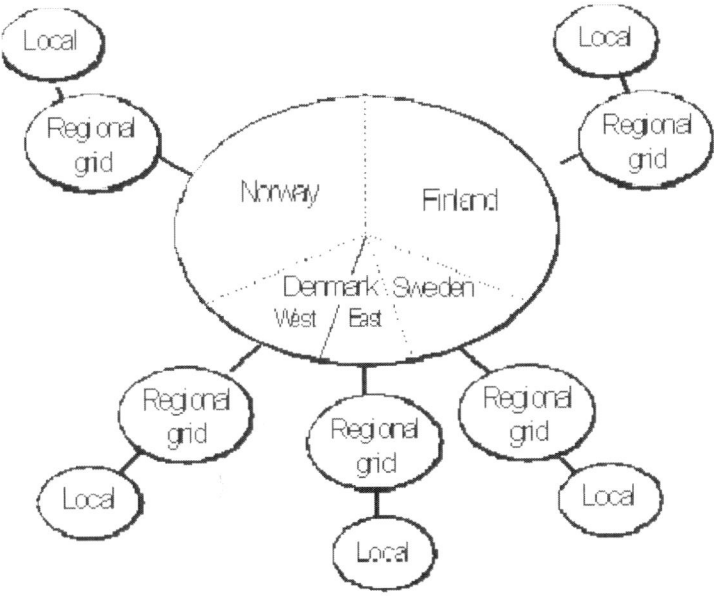

Figure 7.34 Grid structure.

With these levels, there are transmission contracts and tariffs on three levels:

1) Between the central and regional grid and among generators and end users directly connected to the central grid, Statnett has to recover the total central grid cost through these transmission tariffs.

2) Between the regional grid and the distribution grid and between generators and end users directly connected to the regional grid. The regional utility has to recover the total regional grid cost, <u>including the cost of using the central grid</u>, through the T/D contracts.

3) Between the distribution utility and the end users and possible generators connected to the distribution grid. The distribution utility has to recover the total cost of the local grid, <u>including the cost of using the regional and thereby also the central grid</u>.

It follows that in the local point tariff the expenses for using both the regional and central grids are accounted for. Once a customer has paid his point tariff, he is on the national or international market. (from 1996 on the Norwegian and Swedish markets, from 1998 it also included the Finnish market and from 2000 the Danish market as well).

T/D tariffs in the Nordic countries are generally calculated and disseminated once a year. The exception is Norway where the loss element is recalculated every second month.

A brief description of T/D tariffs in the different Nordic countries follows. It will be seen that there are substantial differences between the tariff structures. However, despite these differences, it is possible to accomplish market transactions across national borders. This is one fundamental advantage of a point tariff system. It is possible to live with considerable differences in transmission tariffs and still have a well-functioning market. On the other hand, there are advantages in harmonising the structure, and some work is going on in that direction.

Norway

The Norwegian tariff structure is as follows:

- Operation dependent components:
    - Energy charge (based on marginal losses)
    - Capacity fee
- Operation independent (residual) components:
    - Capacity connection charge
    - Investment charge

The idea is that the operation dependent components contribute to an optimum operation of the system but these components might also have some impact on investment decisions. The operation independent components should ideally not impact the operation. The main purpose is to cover the residual cost without affecting any operational decision.

The operation dependent components of the tariff, the energy and congestion charges, are not sufficient to cover the full cost of the grid. These elements cover 20-35 per cent of the cost, the percentage varies due to the variable spot price which affects the energy charge. The remaining cost must be covered by the so-called operation independent or residual components.

Two elements are included in the operation independent part: a capacity connection charge and an investment charge. First a brief comment on the investment charge, which is by far the least important of the two: The basic idea is that new entrants to the grid, generators as well as consumers, that create a need for new grid investment, can be charged a one-time contribution to cover a part of the grid investment. This should be used especially in cases where new entrants are causing extra costs to the grid that cannot be recovered through their contribution on the other tariff components. This is an interesting principle, but so far it has not been used for the central grid, so there is no practical experience with it. In any case this investment charge will represent a tiny contribution to the total cost of the grid.

This means that the capacity connection charge has to bring in 65-80 per cent of the total cost in the central grid, which makes it a very heavy component in the tariff system.

According to Norwegian regulations the objective of the operation independent components is to cover residual cost. The term operation independent makes it clear that the intention was that this component should not affect the operation. But in certain cases a capacity connection element will have an impact on operational as well as investment decisions. In recent years this tariff element has therefore changed and it is now tied to the expected annual generation (MWh) for the input side. For the consuming side it is still tied to power (maximum load).

Generally there are several partially conflicting targets affecting the design of the residual element:

- *Minimum loss of economic efficiency*. This is what the Norwegian regulations are trying to achieve through the requirement for "operation independency". Another approach leading to the same is the use of Ramsey pricing (see Section 4.4), which means that the least price elastic users should carry most of the residual cost. This is the so-called inverse elasticity rule. Tariffs for interruptible supply, which is used in Norway for electric boilers with oil backup, are based on a simple form of Ramsey pricing. Interruptible load have a lower capacity connection change than ordinary firm consumption.

- *Price equality* is a political target in many cases.

- *Cost responsibility*. A tariff based on the grid cost each customer is responsible for can be regarded as fair. But in many cases it is difficult, if not impossible, to trace the grid cost back individual consumers.

Sweden

The central grid

The point tariff for the national transmission network comprises:

- An annual power fee, which is intended to provide about half of the revenue

- An energy fee, also about half of the revenue

- A one-for-all connection of a new customer involves considerable costs that are not covered by the other two fees

The power fee is latitude-dependent (see Figure 7.35). This is because the prevailing power flow in the network is from north to south. Producers pay more in the north, where there is a surplus of generation, and less in the south where the load centres and export markets are located. Consumers of power will conversely pay more in the south and less in the north. The power fee varies between SEK 1 and SEK 37 per kW and year. The fee is payable for subscribed maximum input and output at each terminal.

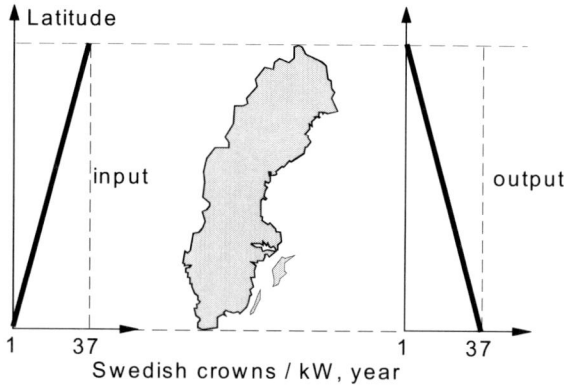

Figure 7.35 The transmission network power fee.

The second element, the energy fee, reflects the short-term marginal cost caused by energy transport. The fee is calculated as the product of the appropriate marginal-loss coefficient for the connection point, the energy input or output, and the current power price. Marginal-loss coefficients for high and low load periods, weekdays and other time periods, have been calculated through load flow calculations for each connection point in the network. These coefficients can be positive or negative as can be seen from the examples in Figure 7.36.

Transmission/Distribution (T/D) tariffs

energy fee =
loss coefficient (+/-)* energy in (+) / out (-)* power price

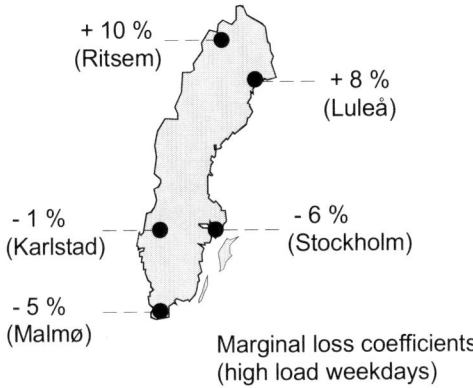

Figure 7.36 The transmission network energy fee.

Svenska Kraftnät is purchasing loss-replacement power on long-term contracts, in contrast to Statnett, which is buying spot power for this purpose.

The third element, the once-for-all connection fee is only be applied in special cases, for example when new connections require considerable investments.

The distribution grid

Distribution grid tariffs in Sweden are not regulated and standardised to the same extent as the Norwegian. The following description applies to Vattenfall Regionnät AB which is the owner of several distribution utilities in Sweden (and some in Finland).

There are different principles for network tariffs in use at Vattenfall Regionnät AB. Since the law admits the use of a geographically dependent and individual tariff for producers the following choice has been made:

- For consumers, an average tariff is used at each distribution level.
- For producers, an individual tariff is based on the actual physical situation.

Tariffs for consumers are not geographically variable due to different costs of the regional distribution network in different parts of the country. On the other hand the tariff for consumers is different in different areas of Sweden due to the geographic dependence that stems from the tariff of Svenska Kraftnät.

Finland and Denmark

Finland and Denmark have recently entered the open market and details concerning T/D tariffs are still under discussion.

7.4.3 Congestion management in the Nordic system

When Norway and Sweden established a common power market in January 1996, the two countries had different practices regarding congestion management in the central grid. Norway dealt with bottlenecks with a capacity fee (or congestion fee) based on a number of price areas. Sweden treated bottlenecks with a buy-back procedure. The two procedures are described in Section 7.2. These two different procedures where maintained in the integrated Nordic market.

Presently, in the Nordic system Finland, Sweden, Denmark East and Denmark West constitute one price area each, whereas Norway is divided into two or more price areas. Congestions between the price areas are handled through the spot market as described in Section 7.2.2.

Within each price area, buy-back procedures are used. In some international literature this is called off merit scheduling, which implies economic compensation to the generators being rescheduled. The buy-back trade implies, in principle, that not only generation is rescheduled but also consumption. In the Nordic system, the balancing market is used for the buy-back procedure. Originally only generators were included in the buy-back process in Sweden. Presently, the consuming side is taking part in balancing market and is therefore also involved in this process.

7.4.4 Congestion management in Europe

As indicated earlier, different terminologies are used concerning congestion managements and system operation.

In this section, concepts and solutions presently discussed in Europe is described. The following description is taken from Nord Pool.

Explicit and implicit capacity auctions – market coupling and market splitting

Explicit capacity auction
Explicit auction is when the transmission capacity on an interconnector is auctioned to the market separately and independently from the marketplaces where electricity is auctioned. Explicit auction is considered as a simple method of handling the capacity on the international interconnections in Europe. The capacity

is normally auctioned in portions through annual, monthly and daily auctions. Since the two commodities transmission capacity and electricity are traded at two separate auctions, there is a lack of information about the prices of the other commodity. This lack of information can result in an inefficient utilization of interconnectors, i.e. less social welfare, less price convergence and more frequent adverse flows.

Implicit capacity auction
ith implicit auction the day-ahead transmission capacity is used to integrate the spot markets in the different bidding areas in order to maximize the overall social welfare in both (or more) markets. The flow on an interconnector is found based on market data from the marketplace/s in the connected markets. Thus the auctioning of transmission capacity is included (implicitly) in the auctions of electricity in the market. In implicit auctions the transmission capacity between bidding areas (price areas) is made available to the spot price mechanism in addition to bid/offers per area, thus the resulting prices per area reflect both the cost of energy in each internal bidding area (price area) and the cost of congestion. Implicit auctions ensure that electricity flows from the surplus areas (low price areas) towards the deficit areas (high price areas) thus also leading to price convergence. Implicit auction signifies the concept used for both 'market coupling' and 'market splitting'. There is not necessarily any difference in the calculation algorithms or principals used for market coupling and market splitting. What differentiates market coupling from market splitting is how the algorithm is operated and owned, and which results from the central calculation the local markets use subsequently.

Market splitting
In market splitting the implicit auction of transmission capacity is handled within the day ahead electricity auction by one single power exchange. In the Nordic region Nord Pool Spot performs market splitting since the transmission capacity between the Nordic bidding areas is handled implicitly in the price and bid matching calculation performed at Nord Pool Spot. Sometimes the transmission capacity between the internal bidding areas is not enough to get a complete convergence of price, and the result is that there are different prices in different bidding areas. Thus the term 'market splitting' refers to the fact that the limited transmission capacity leads to a split between to market areas.

Market coupling
In market coupling the implicit auction is organized in cooperation between two or more power exchanges. Thus the term refers to the coupling of two or more power exchanges. Each power exchange submits the necessary market information to a central coupling algorithm. The TSOs provide the transmission capacity available between the market areas. The central calculation delivers both flows between, and prices in all market areas. The flows between the coupled markets are at least used further in the local bid matching of the power exchanges, but it is also possible

that the local market can adapt also the prices and bid results from the central coupling algorithm. It is common to differentiate between price market coupling and tight and loose volume market coupling.

Price coupling

In price coupling all market data and all market rules of the coupled markets are included in the central market coupling calculation. The central algorithm determines the prices in the underlying bidding areas, a list of selected block orders for each bidding area and the net positions (or flows) between the bidding areas. All this information is then adapted by each power exchange which then merely calculates each participant program based on the prices and block bid selection delivered by the central coupling unit.

Tight volume coupling

In tight volume coupling the same input is required from the power exchanges as for price coupling. However, only the determined flows between each exchange area are adapted by each power exchange which again calculates the prices for the different market areas separately. Since prices are calculated by each power exchange in a second step, volume coupling may result in small adverse flows or price discrepancies due to small differences in matching algorithms, in the implementation of market rules or in the completeness of the market data delivered to the central algorithm. The rationale behind using volume coupling instead of implementing price coupling may be that it is not possible to establish market coupling and at the same time include all market rules and features for all areas at the same time. Volume coupling is by many viewed as a step towards a closer integration of markets on the way towards a common price coupling for larger regions.

Loose volume coupling

As with tight volume coupling, it is only the determined flows (volumes) between each exchange area which are adapted by each power exchange in their local price calculations. However, the more differences there are between the matching algorithms, the less market rules that are implemented in the central algorithm and the less completeness of market data delivered from the power exchanges - the looser becomes the volume coupling. Depending on the 'degree of looseness', loose volume coupling will frequently deliver adverse flows and less convergence of prices between the market areas which again gives a lower gain in social welfare than with tight volume coupling or price coupling.

Market coupling of NPS and EEX through EMCC

On the 9th of November 2009 the market coupling of the Nordic and the German spot market was launched. The market coupling is governed through EMCC - European Market Coupling Company GmbH - which is a joint venture of Nord Pool Spot, EEX Spot , 50Hertz Transmission GmbH (formerly Vattenfall Europe

Transmission), transpower stromübertragungs gmbh (formerly E.ON Netz) and Energinet.dk

EMCC owns a market coupling algorithm which, by calculating the optimal flow on the connections between Denmark and Germany, couples the Nordic and German day ahead markets of Nord Pool Spot and EEX Spot.

The calculation algorithm is physically operated by trading desk personnel of the two power exchanges. NPS and EEX are taking turns of the operation every second month. EMCC is accredited as trading participant at both power exchanges and is entitled by the capacity owners, Transpower, 50Hertz and Energinet.dk, to allocate available cross-border capacities for implicit auctions. Thus the congestion rent from these connections is accumulated in EMCC and subsequently paid to the capacity owners.

The coupling of NPS and EEX is tight volume coupling

By definition, market coupling is the use of so-called implicit auctioning involving two or more power exchanges (PX). Market coupling can in practice be implemented in several ways. The market coupling of NPS and EEX is what is normally referred to as tight volume coupling.

All market rules of the two regions are reflected in the EMCC algorithm and all possible market information from the market places is included in the market coupling calculation.

After gate closure, each power exchange aggregates the bid information for each bidding area and submits these market data in a file to the coupling algorithm of EMCC. The owners of the interconnectors between the market areas put all or part of their capacity at the disposal of EMCC and the capacity values are also sent to the coupling algorithm.

The algorithm uses the economic welfare criterion and on the basis of market coupling capacities and prices, the optimal flow between the market areas is determined. These flows are called market coupling flows (MCF). After the calculation, EMCC submits additional price-independent bids/offers to the power exchanges. These bids and offers reflect the calculated market coupling flow. The exchanges then calculate their own prices taking the bids from EMCC into account.

7.5 Nordic power exchange before 1991

Before the restructuring, there was extensive power exchange between the Nordic countries. NORDEL, which is an institution for cooperation on electricity supply within the area, was established in 1963. The NORDEL cooperation was originally based on the condition that each country should have a satisfactory power and energy balance. If a country was disconnected from its neighbours, it should be able to cover its own demand.

One consequence of this rule was that the whole potential benefit of cooperation never became fully realized. Operational benefit was gained, but not the potential benefit connected with reduced investment. Even though the operational benefit was substantial, there was a large potential for reduced investments in generation capacity caused by the fact that parts of the system was energy constrained (Norway and Sweden) and parts of it was capacity constrained (Finland and Denmark).

The fact that Norway and Sweden can import energy (off-peak) from the others in a dry year and Finland and Denmark can import peak energy from Norway and Sweden, lowers the necessary investments. It was estimated that the generating capacity in the NORDEL area was over-sized corresponding to about 15 TWh/year in around 1990.[18]

7.5.1 Responsibility for operation of the Nordic system

There was no common Nordic operation management. The national operation centres were responsible for operation of the interconnected NORDEL system through bilateral agreements with the neighbouring centres. Responsible for the operation were:

- Vattenfall in Sweden
- Imatran Voima in Finland
- Statkraft in Norway
- Elkraft and Elsam in Denmark

Notice that before the restructuring these companies were responsible for the main grid, and they had a considerable share of the generation. Each national centre was responsible for the technical as well as economic operation of its own system.

The pre-1991 model for power exchange between the Nordic countries was based on import/export monopolies (or exclusive rights tied to individual lines) in all countries. The companies in charge of the commercial transactions were large companies, State owned or with a large share of public ownership.

---

[18] There were also other reasons for over-sizing. One important factor was the obligation of each individual utility to serve its customers. If this is combined with risk aversion, it will inevitably lead to over-sizing. Another important factor was the possibility for a utility to cover all costs by self-determined tariffs. There was no economic risk associated with over-investment.

In 1971 NORDEL issued a recommendation: "Operational Collaboration within NORDEL", that recommended that the power companies should:

- Utilize the "marginal cost principle" in generation planning; let the running costs for the various production units determine the operation.
- Apply comparable methods for calculating the marginal cost.
- Use the calculated power cost as a basis for generation planning and agreements on power exchanges.
- Use equal profit-sharing in bilateral power exchanges.
- Limit the price for power exchanges to a certain maximum amount (75 SEK/MWh) above the seller's production cost.

This recommendation was the framework for the Nordic power exchange in the 1970s and 1980s.

The main objective for the electric power collaboration was to minimize the total production costs in the NORDEL system. This was achieved by constantly trying to ensure that the production units were used in order of increasing cost (merit order loading), irrespective of location. The production in one country would then probably not match the consumption in this country. Part of the difference can be due to firm power exchanges. The remaining differences represent the optimal exchanges of occasional power. Exchange of occasional power was thus the outcome of a process by which the power companies tried to minimize production cost.

The exchanges of occasional power were also used as a mutual aid between the countries in situations of shortage. For example during dry years in a hydro system, rationing of electric power could be avoided by importing thermal power during off-peak periods.

7.5.2 Power exchange in case of disturbances

In case of outages, power exchange provides an opportunity for using reserve power from neighbouring countries to help the country where the disturbance has occurred.

Consider the situation where a nuclear block in Sweden trips out. By automatic regulation (see the sections on frequency regulation and production reserves, Chapter 9) the instantaneous spinning reserve which is to be kept available in each country, is activated.

Then Sweden must replace the lost generation within 15 minutes by starting fast reserve units, for example gas turbines.

As an alternative, Sweden can contact Norway and agree to buy support power from Norway. Such support power could be agreed upon under the NORDEL rules without any prior notice, and hydro generation, which can be quickly increased, could be bought. Support power was priced by adding the maximum profit margin (75 SEK/MWh) to Norway's marginal value.

The Swedish long-term alternative for replacing the loss of the nuclear-power block might be oil-condensing production. As an alternative to this the operation centre in Sweden can contact for example, Denmark and enquire about a normal power exchange whereby the Swedish oil-condensing cost will be compared with the Danish coal generation cost. If the latter value is lower, power exchange would take place.

This was covered by the NORDEL rules.

7.6 References

[1] R. D. Christie, B: Wollenberg, I. Wangensteen: "Transmission Management in the Deregulated Environment". IEEE Proceedings. February 2000.

[2] I. Wangensteen, A. Gjelsvik: "Transmission tariffs based on optimal power flow" SINTEF Energy Research TR A4669, February 1999.

[3] M. Bjørndal : "Topics on Electricity Transmission Pricing". Dr. oecon. thesis, NHH, Bergen 2000.

[4] Allen J. Wood, Bruce F. Wollenberg: "Power Generation, Operation and Control".John Wiley & Sons, 1996.

[5] Samuel S. Oren, Pablo T. Spiller, Pravin Varaiga, Felix Wu: "Nodal prices and Transmission Rights, A Critical Apprisal", Electricity Journal 8, pp. 24 - 35.

[6] T.-O. Berntsen, N. Flatabø, J.A. Foosnæs, A. Johannesen: "Sensitivity signals in detection of network condition and planning of control actions in a power system". CIGRE-IFAC symposium 39-83, Florence 1983, paper 208-03.

[7] O.B. Fosso, A. Johannesen: "Security contrained short-term hydro-thermal scheduling in a flexible market environment". Proc. Stockholm Power Tech, June 18-22, 1995, IEEE.

[8] W.W. Hogan: "Regional transmission organizations: Millennium order on designing market institutions for electric network systems". Harvard University May 2000. http://ksgwww.harvard.edu/people/whogan

[9] Grasto: "Incentive–based regulation of electricity monopolies in Norway". http://webben.nve.no/english/regulation

[10] O. S. Grande, I. Wangensteen: "Overføringsbegrensninger og nettregulering" TR A5232, juli 2000.

[11] O. S. Grande, I. Wangensteen: " Alternative models for congestion management and pricing. Impact on network planning and physical operation ". CIGRE Session 2000, Paris 2000.

[12] R. Christie and I. Wangensteen "The Energy Market in Norway and Sweden: Congestion Management" IEEE Power Engineering Review, May 1998.

[13] NERA: "Efficient investment in electricity networks: An agenda for Deregulation" Report prepared for NVE, August 1994.

[14] Jan-Pierre Mehr: "Structureal aspacts", Lectures at Seminar on Electric Power System Management at ABB T&D Ludviks Swedwn, November 1999.

[15] Ivan Androcec: "Congestion Management in Short Term Scheduling of Power Generation", Master Thesis Zagreb 2005.

[16] Karsten Neuhof: "Market Power in Networks" PhD Thesis, St. John's College, Cambridge, May 2003.

[17] ETSO: "Coordinated use of Power Exchanges for Congestion Management", Final Report 2001.

[18] T. Kristiansen: "Risk Management in Electricity Markets Emphasizing Tramsmission Congestion" Doctoral Thesis NTNU, Trondheim February 2004.

Appendix to chapter 7

Details concerning optimal power flow in Section 7.3:

No restrictions

We start with the extended objective function (Lagrange function) Equation (7.15)

$$L(\mathbf{x},\mathbf{P},\boldsymbol{\lambda}) = f(\mathbf{P}) + \boldsymbol{\lambda}^t \mathbf{g}(\mathbf{x},\mathbf{P}) \tag{1}$$

We use a three-node network (as shown in Figure 7.21) to show how we can derive the equations for optimal load flow. The Lagrange function can be written:

$$\begin{aligned}L(\mathbf{x},\mathbf{P},\boldsymbol{\lambda}) &= f_1(P_1) + f_2(P_2) + f_3(P_3) \\ &+ \lambda_1 g_1(\mathbf{x},\mathbf{P}) + \lambda_2 g_2(\mathbf{x},\mathbf{P}) + \lambda_3 g_3(\mathbf{x},\mathbf{P})\end{aligned} \tag{2}$$

The vectors are:

$$\mathbf{x} = \begin{pmatrix} x_1 \\ x_2 \\ x_3 \end{pmatrix} = \begin{pmatrix} U_1 \\ U_2 \\ U_3 \end{pmatrix} \tag{3}$$

$$\mathbf{P} = \begin{pmatrix} P_1 \\ P_2 \\ P_3 \end{pmatrix} \tag{4}$$

$$\boldsymbol{\lambda} = \begin{pmatrix} \lambda_1 \\ \lambda_2 \\ \lambda_3 \end{pmatrix} \tag{5}$$

The load flow equations in more detail:

$$g_1(\mathbf{x},\mathbf{P}) = -P_1 + \left[(U^2_1 - U_1 U_2)/R_{12} + (U^2_1 - U_1 U_3)/R_{13} \right] = 0$$

$$g_2(\mathbf{x},\mathbf{P}) = -P_2 + \left[(U^2_2 - U_2 U_1)/R_{12} + (U^2_2 - U_2 U_3)/R_{23} \right] = 0 \quad (6)$$

$$g_3(\mathbf{x},\mathbf{P}) = -P_3 + \left[(U^2_3 - U_3 U_1)/R_{13} + (U^2_3 - U_3 U_2)/R_{23} \right] = 0$$

We find the optimum by derivation of L with respect to λ, $\mathbf{x}$ and $\mathbf{P}$ and then find the point where the derivative is zero. We start with λ:

$$\frac{\partial L}{\partial \lambda} = 0 \Rightarrow \mathbf{g}(\mathbf{x},\mathbf{P}) = 0$$

which is the set of load flow equations.

Next derivation with respect to x:

$$\frac{\partial L}{\partial \mathbf{x}} = 0 \Rightarrow$$

$$\frac{\partial L}{\partial \mathbf{x}} = \begin{pmatrix} \lambda_1 \frac{\partial g_1}{\partial U_1} + \lambda_2 \frac{\partial g_2}{\partial U_1} + \lambda_3 \frac{\partial g_3}{\partial U_1} \\ \lambda_1 \frac{\partial g_1}{\partial U_2} + \lambda_2 \frac{\partial g_2}{\partial U_2} + \lambda_3 \frac{\partial g_3}{\partial U_2} \\ \lambda_1 \frac{\partial g_1}{\partial U_3} + \lambda_2 \frac{\partial g_2}{\partial U_3} + \lambda_3 \frac{\partial g_3}{\partial U_3} \end{pmatrix} = 0 \quad (7)$$

We introduce the Jacobi matrix defined as:

$$\frac{\partial \mathbf{g}}{\partial \mathbf{x}} = \begin{pmatrix} \frac{\partial g_1}{\partial x_1}, \frac{\partial g_1}{\partial x_2}, \frac{\partial g_1}{\partial x_3} \\ \frac{\partial g_2}{\partial x_1}, \frac{\partial g_2}{\partial x_2}, \frac{\partial g_2}{\partial x_3} \\ \frac{\partial g_3}{\partial x_1}, \frac{\partial g_3}{\partial x_2}, \frac{\partial g_3}{\partial x_3} \end{pmatrix} \qquad (8)$$

It is then easy to see that transposing the Jacobi matrix and multiplying with the lambda vector leads to Equation (7.18), which means:

$$\frac{\partial L}{\partial \mathbf{x}} = 0 \Rightarrow \left(\frac{\partial \mathbf{g}}{\partial \mathbf{x}}\right)^t \lambda = 0 \qquad (9)$$

Finally derivation with respect to P:

$$\frac{\partial L}{\partial \mathbf{P}} = 0 \Rightarrow$$

$$\begin{pmatrix} \frac{\partial f_1}{\partial P_1} \\ \frac{\partial f_2}{\partial P_2} \\ \frac{\partial f_3}{\partial P_3} \end{pmatrix} + \lambda^t \begin{pmatrix} \frac{\partial g_1}{\partial P_1} + \frac{\partial g_2}{\partial P_1} + \frac{\partial g_3}{\partial P_1} \\ \frac{\partial g_1}{\partial P_2} + \frac{\partial g_2}{\partial P_2} + \frac{\partial g_3}{\partial P_2} \\ \frac{\partial g_1}{\partial P_3} + \frac{\partial g_2}{\partial P_3} + \frac{\partial g_3}{\partial P_3} \end{pmatrix} = 0 \qquad (10)$$

The elements in the first vector above ($\partial f_i / \partial P_i$) are the marginal costs for input nodes and marginal willingness to pay for output nodes. The elements in the last vector are all -1 as we can easily see from Equation (6). So the result is simply:

$$\begin{pmatrix} c_1 \\ c_2 \\ d_3 \end{pmatrix} + \begin{pmatrix} -\lambda_1 \\ -\lambda_2 \\ -\lambda_3 \end{pmatrix} = 0 \qquad (11)$$

We assume nodes 1 and 2 are input nodes, c1 and c2 corresponding marginal generation costs. Node 3 is an output node d3 is the marginal willingness to pay.

Equations (9), (10) and (11) can then be written:

$$\mathbf{g}(\mathbf{x},\mathbf{P}) = 0 \qquad (12)$$

$$\left(\frac{\partial \mathbf{g}}{\partial \mathbf{x}}\right)^t \lambda = 0 \qquad (13)$$

$$\lambda = \begin{bmatrix} \mathbf{c} \\ \mathbf{d} \end{bmatrix} \qquad (14)$$

This is the set of equations we have to solve for an unconstrained OPF.

Introduction of restrictions

In addition to the load flow equations which must always be met, constraints on the transfer and generation capacity with corresponding Lagrange multiplies η and γ are added. Notice that these are inequality constraints, which can written as,

$$\mathbf{h}(\mathbf{x}) \leq 0 \qquad (15)$$

and:

$$0 \leq P_i \leq P_{imax} \qquad (16)$$

In this case the conditions for optimality can be derived from the Karush-Kuhn-Tucker conditions. According to these conditions, all <u>active constraints</u>, i.e. constraints where we turn against the limit, have to be taken into account and the inequality sign is then replaced with an equality sign. Non-active constraints can be omitted.

The Lagrange function can be written:

$$L(\mathbf{x},\mathbf{P}) = f(\mathbf{P}) + \boldsymbol{\lambda}^t \mathbf{g}(\mathbf{x},\mathbf{P}) + \boldsymbol{\eta}^t \mathbf{h}(\mathbf{x}) + \boldsymbol{\gamma}^t (\mathbf{P} - \mathbf{P}_{max}) \quad (17)$$

For a three-node network (as shown in Figure 7.24) this function can be written:

$$\begin{aligned} L(\mathbf{x},\mathbf{P},\lambda) = &f_1(P_1) + f_2(P_2) + f_3(P_3) \\ &+ \lambda_1 g_1(\mathbf{x},\mathbf{P}) + \lambda_2 g_2(\mathbf{x},\mathbf{P}) + \lambda_3 g_3(\mathbf{x},\mathbf{P}) \\ &+ \eta_1 h_1(\mathbf{x}) + \eta_2 h_2(\mathbf{x}) + \eta_3 h_3(\mathbf{x}) \\ &+ \gamma_1(P_1 - P_{1,max}) + \gamma_2(P_2 - P_{2,max}) + \gamma_3(P_3 - P_{3,max}) \end{aligned} \quad (18)$$

We next find the derivative of L with respect to $\boldsymbol{\lambda}$, $\boldsymbol{\eta}$, $\boldsymbol{\gamma}$, x and **P** and find the point where the derivative is zero. Derivation with respect to the Lagrange multipliers $\boldsymbol{\lambda}$, $\boldsymbol{\eta}$ and $\boldsymbol{\gamma}$ leads to the following equations:

$$\mathbf{g}(\mathbf{x},\mathbf{P}) = 0$$
$$\mathbf{h}(\mathbf{x}) = 0 \quad (19)$$
$$(P - Pmax) = 0$$

Notice that inequality is here replaced by equality. In addition all non-active constraints can be left out.

Derivation with respect to the state vector x:

$$\frac{\partial L}{\partial \mathbf{x}} = \begin{pmatrix} \lambda_1 \frac{\partial g_1}{\partial U_1} + \lambda_2 \frac{\partial g_2}{\partial U_1} + \lambda_3 \frac{\partial g_3}{\partial U_1} \\ \lambda_1 \frac{\partial g_1}{\partial U_2} + \lambda_2 \frac{\partial g_2}{\partial U_2} + \lambda_3 \frac{\partial g_3}{\partial U_2} \\ \lambda_1 \frac{\partial g_1}{\partial U_3} + \lambda_2 \frac{\partial g_2}{\partial U_3} + \lambda_3 \frac{\partial g_3}{\partial U_3} \end{pmatrix} + \begin{pmatrix} \eta_1 \frac{\partial h_1}{\partial U_1} + \eta_2 \frac{\partial h_2}{\partial U_1} + \eta_3 \frac{\partial h_3}{\partial U_1} \\ \eta_1 \frac{\partial h_1}{\partial U_2} + \eta_2 \frac{\partial h_2}{\partial U_2} + \eta_3 \frac{\partial h_3}{\partial U_2} \\ \eta_1 \frac{\partial h_1}{\partial U_3} + \eta_2 \frac{\partial h_2}{\partial U_3} + \eta_3 \frac{\partial h_3}{\partial U_3} \end{pmatrix} = 0 \quad (20)$$

This equation can be written:

$$\left(\frac{\partial \mathbf{g}}{\partial \mathbf{x}}\right)^t \boldsymbol{\lambda} + \left(\frac{\partial \mathbf{h}}{\partial \mathbf{x}}\right)^t \boldsymbol{\eta} = 0 \quad (21)$$

Finally derivation with respect to P:

$$\frac{\partial L}{\partial \mathbf{P}} = 0 \Rightarrow$$

(22)

$$\begin{pmatrix} \frac{\partial f_1}{\partial P_1} \\ \frac{\partial f_2}{\partial P_2} \\ \frac{\partial f_3}{\partial P_3} \end{pmatrix} + \boldsymbol{\lambda}^t \begin{pmatrix} \frac{\partial g_1}{\partial P_1} + \frac{\partial g_2}{\partial P_1} + \frac{\partial g_3}{\partial P_1} \\ \frac{\partial g_1}{\partial P_2} + \frac{\partial g_2}{\partial P_2} + \frac{\partial g_3}{\partial P_2} \\ \frac{\partial g_1}{\partial P_3} + \frac{\partial g_2}{\partial P_3} + \frac{\partial g_3}{\partial P_3} \end{pmatrix} + \boldsymbol{\gamma}^t \begin{pmatrix} 1 \\ 1 \\ 1 \end{pmatrix} = 0 \quad (22)$$

$$\begin{pmatrix} c_1 \\ c_2 \\ d_3 \end{pmatrix} + \begin{pmatrix} -\lambda_1 \\ -\lambda_2 \\ -\lambda_3 \end{pmatrix} + \begin{pmatrix} \gamma_1 \\ \gamma_2 \\ \gamma_3 \end{pmatrix} = 0 \Rightarrow \quad (23)$$

$$\boldsymbol{\lambda} = \begin{bmatrix} \mathbf{c} + \boldsymbol{\gamma} \\ \mathbf{d} \end{bmatrix} \quad (24)$$

The full set of equations for the constrained case:

$$g(x, P) = 0$$

$$h(x) = 0$$

$$(P - P_{max}) = 0$$

$$\left(\frac{\partial g}{\partial x}\right)^t \lambda + \left(\frac{\partial h}{\partial x}\right)^t \eta = 0$$

$$\lambda = \begin{bmatrix} c + \gamma \\ d \end{bmatrix}$$

h(x) represents active restrictions only, i.e. cases where *h(x)* = 0. If *h(x)* < 0, the corresponding equation is not included. The same applies to restrictions on generating capacity.

Nodal prices and marginal transmission losses

The link between optimal nodal prices and marginal transmission losses can be illustrated by looking at an arbitrary line connecting two nodes i and j shown in Figure 1.

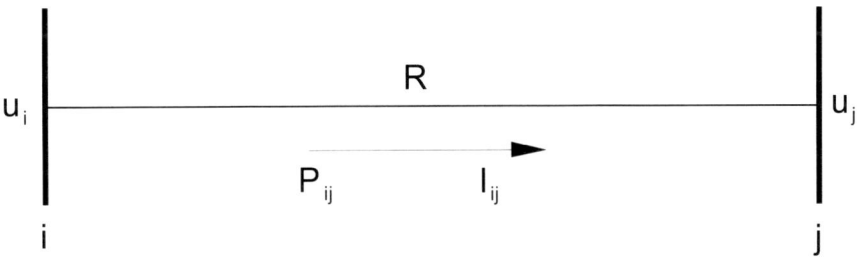

Figure 1. Transmission line connecting two nodes.

Seen from node i, the current is

$$I_{12} = (U_1 - U_2)\frac{1}{R} \quad (25)$$

and the power is

$$P_{12} = U_1 I_{12} \qquad (26)$$

The transfer losses are

$$Loss_{12} = R I_{12}^2 = R \left(\frac{P_{12}}{U_1} \right)^2 \qquad (27)$$

We can now find the marginal loss for transfer from i to j seen from node i:

$$\frac{\partial Loss_{12}}{\partial P_{12}} = 2 \frac{R \cdot P_{12}}{U_i^2} = 2 \left(1 - \frac{U_2}{U_1} \right) \qquad (28)$$

This means that if we transmit a small quantity ΔP_i from node i, and the corresponding quantity arriving at node j is ΔP_j, then ρ_{ij} is given by:

$$\rho_{12} = \frac{\Delta P_1 - \Delta P_2}{\Delta P_1} = 2 \left(1 - \frac{U_2}{U_1} \right) \qquad (29)$$

A reformulation of this equation gives:

$$\frac{U_2}{U_1} = 1 - \frac{\rho_{12}}{2} \qquad (30)$$

We now go back to the load flow equations (6):

$$g_1(\mathbf{x},\mathbf{P}) = -P_1 + \left[(U^2_1 - U_1 U_2)/R\right] = 0$$

$$g_2(\mathbf{x},\mathbf{P}) = -P_2 + \left[(U^2_2 - U_2 U_1)/R\right] = 0 \qquad (31)$$

This gives the following set of equations for the lambdas:

$$\left(\frac{\partial \mathbf{g}}{\partial \mathbf{x}}\right)^t \lambda = 0 \Rightarrow \qquad (32)$$

$$\lambda_1 \left[(2U_1 - U_2)/R\right] + \lambda_2 \left[-U_2/R\right] = 0$$

$$\lambda_1 \left[-U_1/R\right] + \lambda_2 \left[(2U_2 - U_1)/R\right] = 0 \qquad (33)$$

We look at the last of these two equations. From that we find the following relationship between the lambdas:

$$\frac{\lambda_1}{\lambda_2} = 2\frac{U_2}{U_1} - 1 \qquad (34)$$

We combine this with equation (29) and find that:

$$\frac{\lambda_1}{\lambda_2} = 1 - \rho_{12}$$

or:

$$\rho_{12} = 1 - \frac{\lambda_1}{\lambda_2} = \frac{\lambda_2 - \lambda_1}{\lambda_2} \qquad (35)$$

From this we can draw a more general conclusion for the relationship between nodal prices and marginal losses:

$$\frac{\lambda_i}{\lambda_j} = 1 - \rho_{ij} \qquad (36)$$

Solution of the three-node unconstrained case

We start with the load flow equations:

$$g(x, P) = 0 \qquad (37)$$

The bus admittance matrix is:

$$Y = \begin{bmatrix} 0,3 & -0,1 & -0,2 \\ -0,1 & 0,15 & -0,05 \\ -0,2 & -0,05 & 0,25 \end{bmatrix} \qquad (38)$$

and the current vector.

$$\begin{bmatrix} I_1 \\ I_2 \\ I_3 \end{bmatrix} = Yx = \begin{bmatrix} 0,3U_1 & -0,1U_2 & -0,2U_3 \\ -0,1U_1 & 0,15U_2 & -0,05U_3 \\ -0,2U_1 & -0,05U_2 & 0,25U_3 \end{bmatrix} \qquad (39)$$

Input in each node (which is the control vector) is derived by multiplying current and voltage.

$$\begin{bmatrix} P_1 \\ P_2 \\ P_3 \end{bmatrix} = \begin{bmatrix} I_1 U_1 \\ I_2 U_2 \\ I_3 U_3 \end{bmatrix} = \begin{bmatrix} 0,3U_1^2 & -0,1U_2 U_1 & -0,2U_3 U_1 \\ -0,1U_1 U_2 & 0,15U_2^2 & -0,05U_3 U_2 \\ -0,2U_1 U_3 & -0,05U_2 U_3 & 0,25U_3^2 \end{bmatrix} \qquad (40)$$

The load flow equations

$$g(\mathbf{x}, \mathbf{P}) = 0$$

can then be written out.

$$0{,}3U_1^2 - 0{,}1U_2U_1 - 0{,}2U_3U_1 - P_1 = 0 \qquad (41) \quad (a)$$

$$-0{,}1U_1U_2 + 0{,}15U_2^2 - 0{,}05U_3U_2 - P_2 = 0 \qquad (b)$$

$$-0{,}2U_1U_3 - 0{,}05U_2U_3 + 0{,}25U_3^2 - P_3 = 0 \qquad (c)$$

In order to solve this set of equations, we first chose the voltage in one node, here node 3.

$$U_3 = 60 \text{ kV}$$

It is further given that P3 = -30 MW. Since there will be no congestion, it is obvious from the marginal costs, that generator 1 will generate at maximum (20 MW) and generator 2 will be the slack or swing producer.

Equation (39) then leads to the following set of solvable equations.

$$0{,}3U_1^2 - 0{,}1U_2U_1 - 0{,}2 \cdot 60 \cdot U_1 - 20_1 = 0 \qquad (42) \quad (a)$$

$$-0{,}1U_1U_2 + 0{,}15U_2^2 - 0{,}05 \cdot 60 \cdot U_2 - P_2 = 0 \qquad (b)$$

$$-0{,}2U_1 \cdot 60 - 0{,}05U_2 \cdot 60 + 0{,}25 \cdot 60^2 - (-30) = 0 \qquad (c)$$

Equation c) can be used to find U2 as a function of U1

$$U_2 = 310 - 4U_1 \qquad (43)$$

Substitution for U2 in Equation (40) a) gives,

$$20 = U_1[0{,}2(U_1 - 60) + 0{,}1(U_1 - 310 + 4U_1)] \quad (44)$$

and this leads to the equation

$$0{,}7U_1^2 - 43U_1 - 20 = 0 \quad (45)$$

with the solution:

$$U_1 = 61{,}89 \text{ kV}$$

Equation (41) then gives:

$$U_2 = 62{,}44 \text{ kV}$$

Then we have the complete state vector

$$\mathbf{x} = \begin{bmatrix} U_1 \\ U_2 \\ U_3 \end{bmatrix} = \begin{bmatrix} 61.89 \\ 62.44 \\ 60.00 \end{bmatrix} \quad [kV]$$

Using Equation (40) b) we can find the unknown generation from unit 2:

$$P_2 = 11{,}05 \text{ MW}$$

The sum of input in all nodes gives the grid losses:

$$P_{loss} = 1{,}05 \text{ MW}$$

Next we find the lambdas, i.e. the nodal prices. This is done by first finding the derivatives of g with respect to x. We start with x as described. We notice that U_3 is fixed which means that only derivation with respect to U1 and U2 is necessary:

$$\frac{\partial \mathbf{g}}{\partial \mathbf{x}} = \begin{bmatrix} 0.6\,U_1 - 0.1\,U_2 - 0.2\,U_3 & -0.1\,U_1 \\ -0.1\,U_2 & 0.3\,U_2 - 0.1\,U_1 - 0.05\,U_3 \\ -0.2\,U_3 & -0.05\,U_3 \end{bmatrix} \qquad (46)$$

Numerical values of U1, U2 and U3 give:

$$\frac{\partial \mathbf{g}}{\partial \mathbf{x}} = \begin{bmatrix} 18.89 & -6.189 \\ -6.244 & 9.543 \\ -12 & -3 \end{bmatrix} \qquad (47)$$

This matrix is transposed and multiplied with the vector λ:

$$\left(\frac{\partial \mathbf{g}}{\partial \mathbf{x}}\right)^t \lambda = \begin{bmatrix} 18.89\lambda_1 - 6.244\lambda_2 - 12.00\lambda_3 \\ -6.189\lambda_1 + 9.543\lambda_2 - 3.00\lambda_3 \end{bmatrix} = \begin{bmatrix} 0 \\ 0 \end{bmatrix} \qquad (48)$$

We now have 2 equation, but 3 unknowns. The third equation is given by the marginal cost in node 2 where $\lambda_2 = 14$ øre/kWh. We can then find the two remaining unknowns.

$\lambda 1 = 14.247$ øre/kWh
$\lambda 3 = 15.148$ øre/kWh

The nodal prices are then:

p1 = $\lambda 1$ = c1 + γ1 = 12 + 2.247 øre/kWh
p2 = $\lambda 2$ = c2 = 14 øre/kWh
p3 = $\lambda 3$ = d3 = 15.148 øre/kWh

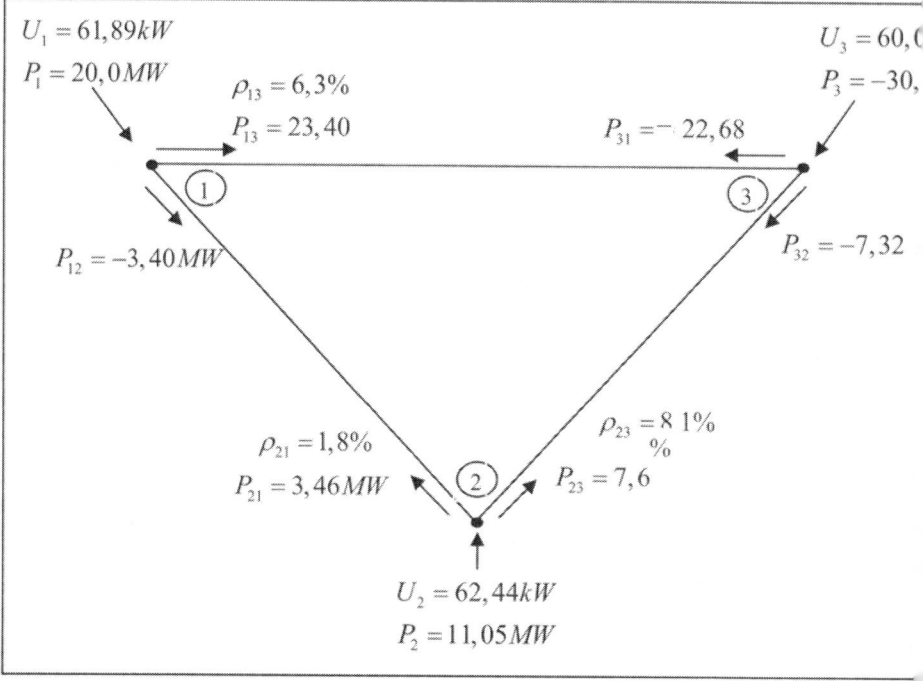

Figure 2. Uncongested 3 node network

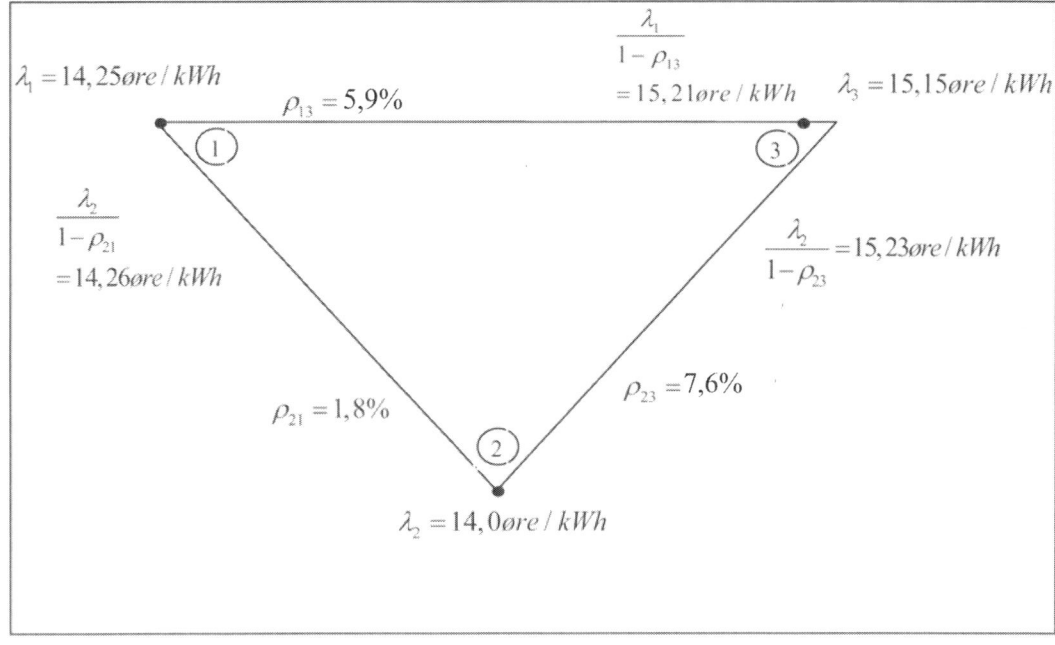

Figure 3. Marginal losses and nodal prices.

Congested three-node example

We introduce a congestion between node 1 and 3. Maximum current is set to 350 A which means that

$$I_{1,3} = \frac{1}{R_{31}}(U_1 - U_3) \le I_{1,3\max} \qquad (49)$$

or

$$U_1 - U_3 \le R_{31} \cdot I_{3\max} = \Delta U_{\lim} = 1{,}75 \qquad (50)$$

Chapter 7: Grid access

$$\mathbf{h}(\mathbf{x}) = \begin{bmatrix} |0.2(U_1-U_3)| - I_{1,3\max} \\ |0.1(U_1-U_2)| - I_{1,2\max} \\ |0.05(U_2-U_3)| - I_{2,3\max} \end{bmatrix}$$

$$\frac{\partial \mathbf{h}}{\partial \mathbf{x}} = \begin{bmatrix} 0.2 & 0 \\ 0.1 & -0.1 \\ 0 & 0.05 \end{bmatrix} \qquad (51)$$

In our case the only active transfer restriction occurs at the line from node 1 to node 3. This leads to the following formulation of the restrictions on transfer capacity:

$$\left(\frac{\partial \mathbf{h}}{\partial \mathbf{x}}\right)' \boldsymbol{\eta} = \begin{bmatrix} 0.2\eta \\ 0 \end{bmatrix} \qquad (52)$$

And the set of equations becomes

$$\begin{aligned} 0.3U_1^2 - 0.1U_2U_1 - 0.2 \cdot 60 \cdot U_1 - P_1 &= 0 \\ -0.1U_1U_2 + 0.15U_2^2 - 0.05U_2 \cdot 60 - P_2 &= 0 \\ -0.2U_1 \cdot 60 - 0.05U_2 \cdot 60 + 0.25 \cdot 60^2 - (-30) &= 0 \end{aligned} \qquad \text{(a)}$$

$$U_1 - 60 = 1.75 \qquad \text{(b)}$$

$$\begin{aligned} \lambda_1(0.6U_1 - 0.1U_2 - 0.2 \cdot 60) - 0.1U_2\lambda_2 - 0.2 \cdot 60\lambda_3 + 0.2\eta &= 0 \\ -0.1U_1\lambda_1 + \lambda_2(-0.1U_1 + 0.3U_2 - 0.05 \cdot 60) - 0.05 \cdot 60\lambda_3 &= 0 \end{aligned} \qquad \text{(c)}$$

$$\begin{aligned} \lambda_1 &= 12 \\ \lambda_2 &= 14 \end{aligned} \qquad \text{(d)}$$

We have 8 equations and 8 unknowns and we get the following solution:

U1 = 61.75 kV
U2 = 63.00 kV
λ1 = 12 øre/kWh
λ2 = 14 øre/kWh
λ3 = 20.683 øre/kWh
P1 = 13.894 MW
P2 = 17.325 MW
η = 5.57 kr/Ah
Losses: 1.219 MW.

If the constraint is defined as maximum power transmitted from 1 to 3 referred to 3, we have $P_{13\max} = I_{3\max} \cdot U_3$ and find similarly that

$$\eta_{kWh} = \frac{\partial f}{\partial P_{13\max}} = 557/60 \, \text{øre/Ah} \times \text{kV} = 9,283 \, \text{øre/kWh}$$

8 Ancillary services

8.1 Purpose and definition

8.1.1 Quality of supply

Ancillary services are services that are fundamental for the quality of a power system. By quality we mean: security of supply, frequency stability, voltage level and voltage stability. These are quality characteristics that are important for the end users.

All these characteristics will be the same (<u>collective</u>) for a set of consumers. All consumers within a synchronous grid receive the same frequency. The voltage quality and the quality of supply will also be the same for several consumers.

The fact that consumers receive the same quality, does not mean that all consumers of electricity have the same quality <u>requirement</u>. The requirement can be different. The System Operator has to make decisions concerning the quality and that will have impact on the need for ancillary services.

In addition to this collective aspect, a special features of electricity is that quantity and quality are disconnected. Individual consumers can buy individual quantities of electricity but not individual qualities. For most other commodities quantities and qualities are tied together in one single product.

Collective or public goods are goods that are nonexclusive. Once the goods are produced, they provide benefit to a group of consumers. Classical examples of collective goods are police and military defence. It is generally acknowledged that a free market alone cannot provide such goods.[1] They must be provided on a collective basis. That means we need a responsible institution, in the case of electricity a System Operator (SO), to take necessary actions. In Norway, Statnett is the System Operator and also responsible for operation of the transmission grid, a Transmission System Operator (TSO).

How quality of supply is linked to the availability of capacity reserves or more generally to the availability of Ancillary Services has been discussed in several papers and reports (see for example [2] and [6]). Without going into that

---

[1] We refer to [4] where it is stated that: "It is virtually impossible to get consumers to reveal their preferences regarding collective goods because rational consumers will try to become free-riders, each understating his demand in the hope of avoiding his share of the cost without affecting the quantity he obtains. Consequently such products cannot be marketed in the conventional way and we cannot use market prices to value them.

discussion here, we can conclude it is generally accepted that such services are fundamental for the quality of supply in a power system. The SO is responsible for the provision of such services.

What is defined as Ancillary Services differs from system to system depending on physical characteristics and to some extent institutional characteristics. Normally it includes *reserves*, in principle reserves for active as well as reactive power. These reserves can be on the generating as well as the consumption side. They can be activated by automatic or manual control and the response time can span from seconds to hours. Normally *reactive power* (not only reactive reserves, but running reactive generation) is included as an Ancillary Service. In a power system dominated by thermal generation, *black start capability* is normally defined as an ancillary service. In order to recover from a blackout, black start capability for at least a few plants in the affected area can be vital. In a hydro system, black start capability is normally included in all plants. Systems for load shedding and generation shedding (activated by extreme frequency deviations) can also be included. In some cases *grid losses* are also included in what is defined as an ancillary service.

The System Operator is normally in a single buyer or a *monopsonistic* position. Generators, consumers or the grid company can produce the services, but the SO is the only buyer. But there are also examples of systems where the System Operator only defines the requirement for Ancillary Services and puts obligations on the participants in the market, in particular on the generating companies, to cover that requirement. Each generator can for instance be instructed (by the SO) to provide capacity reserves corresponding to a certain percentage of the running generation. It is then left to the generating company to self-provide or to buy that on the market. Thereby it is possible to create a market (secondary market) for Ancillary services. This type of arrangement is in operation in some states in the US.

8.2 Classification

Table 8.1. Classification of ancillary services

Active/reactive	Status for reserves	Reserve readiness	Control	Provider
Active Power Reactive Power	In stand-by mode (ready to be activated) In activated mode (called up and participating)	Spinning Hot stand-by Cold stand-by	Automatic Manual	Supply side Demand side
Black start capability Optimization/scheduling/despatch System protection Load following Grid losses				

Table 8.2 Ancillary services in a system dominated by thermal generation

Type		Control (activation)	Time response
Active reserves	Primary reserve	Automatic: frequency	Seconds
	Secondary reserve	Automatic: ACE* Manual	A few minutes 10 – 30 minutes
	Tertiary reserve		
	Load shedding	Automatic: frequency	A few minutes
	Production tripping	Automatic: frequency	A few minutes
Reactive reserves		Automatic: voltage control Manual	Minutes
Reactive generation			
Black start capability			

* ACE – Area Control Error.

Primary reserves are made up of available capacity in units already in operation (spinning reserve). The reserves are automatically activated by frequency deviations.

The Secondary reserve is also normally spinning reserve. It is not normal to include start/stop procedures in the automatic control functions, even if fast starting gas turbines can be used as a part of the Secondary control. Usually, a number of predetermined power plants with good control capabilities are used for this purpose. These power plants are directly linked to the national or regional control centre.

The main criterion for activation of the Secondary control is the Area Control Error (ACE)[2] and/or frequency deviation/time deviation. Producers with pump, or in-storage power plants available, will usually utilize these for the Secondary control since the hydro plants are excellent control sources with hardly any costs.

One important issue here is that pump storage power plants are not adjustable when they are in pump mode. This means that in low load periods when the plant pumps water (night/weekend), other units must contribute to the Secondary control.

---

[2] The Area Control Error (ACE) is defined as the deviation from the scheduled exchange with neighbouring control areas, adjusted for frequency deviation. The use ACE as a criterion for Secondary control makes it possible to regain both normal frequency and scheduled exchange through the Secondary control. More about that in section 8.4.

Table 8.3. Ancillary services in the Norwegian system

Type		Control (activation)	Time response
Balancing reserves	Primary reserve	Automatic: frequency 1) Frequency control reserves 2) Contingency reserves	Seconds
	Secondary reserve	Fast manual reserves	
	Tertiary reserve	Manual Reserve Option Market (ROM) Balancing Market (MB)	15 minutes
System Services	Reactive power	Automatic: voltage control Manual	Minutes
	Load Following	Manual	A few minutes
	System protection :		
	Grid splitting	Manual	A few minutes
	Load shedding	Automatic: frequency	A few minutes
	Production tripping	Automatic: frequency	A few minutes

In the Nordic system (as in thermal systems) we have primary reserve based on automatic control (frequency control). It is divided in two categories: 1) Frequency control reserves and 2) Contingency reserves, see Table 8.3. But in contrast to most thermal systems, our secondary reserve is at the time being, manual. But there are plans to introduce automatic secondary control.

The Norwegian System Operator, divides System Services in two broad categories:

- Balancing services
- System services

as shown in Table 8.3 .

8.2.1 Balancing services

The balancing services are in the traditional way divided into primary, secondary and tertiary reserves. The tertiary reserves are handled by the Balancing Market (BM) and the recently established Reserves Options Market (ROM). We return to that in more detail later.

The Secondary reserve is presently manually activated (by phone), and must be operational within 15 minutes. Automatic Secondary control in the Nordic system has been considered as a means to handle the more rapid generation transients that are foreseen in the future, but it has not been introduced so far.

8.2.2 System services

Reactive power/Voltage control

The reactive power is defined as an Ancillary Service. This holds for both the running reactive production, and the reactive reserves that are held as a backup. Correct allocation of the reactive power generation is very important in order to obtain a satisfactory voltage quality and to reduce losses.

It is sometimes difficult to separate Ancillary Services from services or commodities that can be traded in a normal open market. One example is reactive power. It has been proposed that reactive power can be traded in basically the same way as active power. However, one of the problems with reactive power in that context is that it closely linked to voltage. If reactive power is injected into a certain node, the voltage level in that node will rise. That makes open trade extremely difficult in practice. The largest consumers of reactive power are the grid companies, which have substantial reactive losses. Some reactive power is also consumed by the end users, but little compared to the grid companies. The grid company and the SO are therefore normally given the responsibility of supplying the grid and end users with reactive power. That is a natural solution in a situation where system operation and transmission grid operation is integrated in one TSO institution.

Before the deregulation, Ancillary Services were provided by the largest generators, (in Norway most notably by Statkraft) without any explicit payment. It was regarded as a national responsibility of a state owned (or more generally a publicly owned) utility to contribute to a determined quality of supply. The cost could be covered through the customer price.

Load following

Load following includes a detailed monitoring and adjustments of the balance in order to maintain good frequency quality during periods of rapid load changes. Especially during the morning hours there is a need for more frequent adjustment of the balance than what is included in the hour-by-hour schedule established as a result of the day-ahead market.

System protection

Grid splitting

In certain emergency situation it can be possible to prevent a complete system break down if the grid is split.

Load shedding

A substantial decrease in frequency (below 48.7 Hz) activates the Load shedding. This means that it will only happen when the frequency decreases more than the threshold that triggers the operational disturbance reserves. Therefore, load shedding occurs very seldom.

Production tripping

Production tripping is used as a system protection to limit operational disturbances. Production tripping can, for example, be activated due to outages of critical components or from overloaded sections in the grid. Production tripping can also be activated by development of a too high frequency.

8.3 Reserve requirement

As a general (slightly simplified) rule, the requirement for short-term (spinning) reserve in a generating system is set by the largest unit in the system. The system should be able to cover demand if an outage of the largest unit should occur, the so-called (n-1) criterion.

It follows from this that the total short-term reserve requirement for each subsystem is reduced when several subsystems are integrated into one. It is also obvious that several small subsystems have more to gain with respect to reserve requirement than large subsystems. This depends on the fact that in a small system, the largest unit is probably larger in relation to total system capacity than in a large system.

Presently reserve requirements in the Nordic area are defined (by the System Operators) as fixed. There is no flexibility. It is discussed, however, to use a flexible requirement making it possible to handle ancillary services on a cost-benefit basis, see [14].

8.4 Automatic control

Primary, and in most cases secondary reserves are controlled by automatic control based on frequency. If the balance in the system suddenly changes, the primary control brings it back to a stable state with a stationary frequency deviation. Figure 8.1 shows a case.

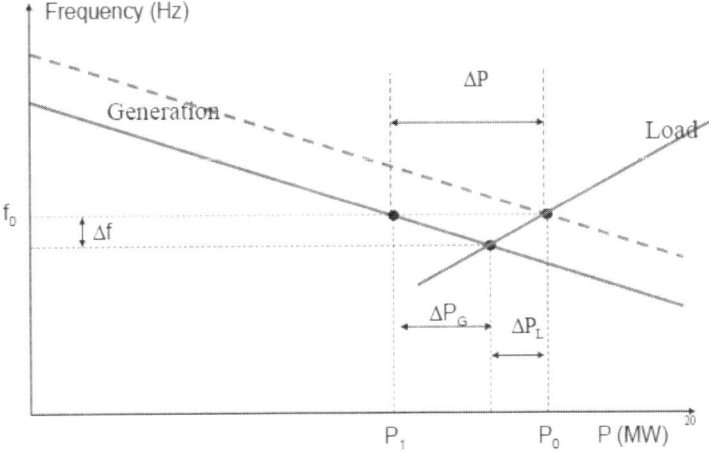

Figure 8.1. Frequency control

We start with a frequency f_0 and a balanced load/generation P_0. Then we assume a sudden imbalance occurs due to the loss of a generating unit with capacity ΔP. That will cause a reduction of the load and an increase in generation (from the remaining units) at the same time as the frequency drops as shown in the figure.

The load and the generating level both depend on frequency but for different reasons. The load will drop with decreasing frequency due to the fact that some types of load, for instance electric motors, will draw less power with reduced velocity. On the generating side the load frequency characteristic depends on the turbine controllers. One adjustable parameter of a controller is called *droop* (NO.:*statikk*). That is defined as:

$$s = -\frac{\Delta f/f_n}{\Delta P/P_n} \cdot 100\% \qquad (8.1)$$

Where P_n is the rated capacity of the unit, f_n is the rated frequency (50 Hz) and ΔP is the additional output cased by frequency change Δf. The formula indicates that a drop in frequency gives increased output. The corresponding *frequency bias* is defined as:

$$R = \Delta P/\Delta f \; [MW/Hz] \qquad (8.2)$$

For one unit the frequency bias will be:

$$R_{unit} = -2 P_n/s \; [MW/Hz] \qquad (8.3)$$

and for a control area with several units the total frequency bias from the generating side is the sum of each unit's contribution:

$$R_G = \sum R_{unit} = \frac{\Delta P_G}{\Delta f} [MW/Hz] \qquad (8.4)$$

As mentioned, there is also a contribution from the load:

$$R_L = \frac{\Delta P_L}{\Delta f} [MW/Hz] \qquad (8.5)$$

Notice that R_G is negative and R_L is positive.

Automatic control

The total frequency bias for the control area is:

$$R = R_G - R_L = \frac{\Delta P_G - \Delta P_L}{\Delta f} = \frac{\Delta P}{\Delta f} \qquad (8.6)$$

The total frequency bias is the sum of the generation and the load side.

That is also evident from Figure 8.1. The figure depicts supply and demand of electricity as a function of frequency. We notice that supply drops with increasing frequency and demand increases with increasing frequency. It is the opposite of what we see if we look on <u>price</u> dependency for supply and demand.

The loss of one generating unit is represented by a shift in the supply curve. The supply curve moves ΔP to the left, from the dotted curve to the solid. The resulting imbalance is compensated by an increase ΔP_G of generation and a decrease ΔP_L of the load at the same time as the frequency drops Δf. So the power balance is re-established with a permanent frequency deviation. When we move to the secondary control, the frequency can be brought back to its normal value, for instance by bringing more generation in.

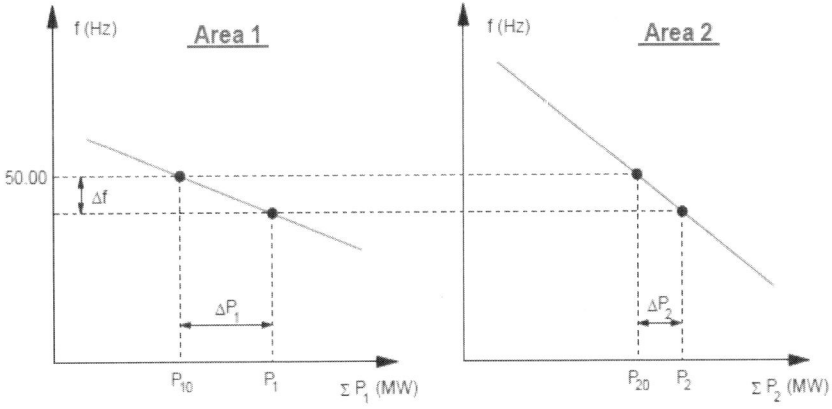

Figure 8.2. Two control areas.

We next look at two control areas with different frequency bias R_1 and R_2 as indicated in the figure above. We will then introduce a new concept: *The Area*

Control Error, ACE (Norw.: innstillingsfeil) and show how that can be used for secondary control.

The Area Control Error is defined as:

$$I_i = \Delta P_{xi} + R_i \cdot \Delta f \qquad (8.7)$$

Where:

- ΔP_{xi} is the deviation in power transfer to area i from other area(s).
- R_i is frequency bias in area i.
- Δf is the frequency deviation (same in the two areas)

The ACE gives an indication concerning which area is causing an imbalance and is therefore suitable for secondary control.

Consider one case where we have an outage of a generating unit in area 1 and lose ΔP_f. Primary control in both areas will be active and the resulting frequency deviation will be:

$$\Delta f = \frac{\Delta P_f}{R_1 + R_2} \qquad (8.8)$$

In area 1 the activated primary reserves is $\Delta P_1 = R_1 \Delta f$, and in area 2 it is $\Delta P_2 = R_2 \Delta f$. That leads to increased transfer to area 1: $\Delta P_{x1} = -\Delta P_{x2} = R_2 \Delta f$.

We next look at the ACE in the two areas:

Area 1.:
$$\begin{aligned} I_1 &= \Delta P_{x1} + R_1 \cdot \Delta f \\ &= R_2 \cdot \Delta f + R_1 \cdot \Delta f \\ &= (R_2 + R_1) \cdot \Delta f = -\Delta P_f \end{aligned} \qquad (8.9)$$

Area 2:
$$\begin{aligned} I_2 &= \Delta P_{x2} + R_2 \cdot \Delta f \\ &= -R_2 \Delta f + R_2 \cdot \Delta f = 0 \end{aligned} \qquad (8.10)$$

The ACE for area 1 indicates a deficit of ΔP_f while ACE for area 2 indicates zero. Secondary regulation based on ACE will therefore bring in ΔP_f new generation in area 1 and no new generation in area 2. That will re-establish the original balance in both areas and bring transfer between the areas back to zero.

8.5 Balancing

After the day-ahead trade is settled and the production plans are decided, imbalances can occur due to changes in the predicted consumption and production, outages of transmission lines, faults in the system or outages of plants. Balancing within the hour is therefore necessary to be able to maintain a satisfying system quality. This is done through intraday trade or through use of the reserves. The following description is tied to the Nordic market, but it is based general principles.

Intraday trade makes it possible for balance responsible entities to adjust their plans closer to real time. Intraday trade (Elbas) was implemented in Norway in 2009 by NordPool, and it is now available in all the Nordic countries. The participants are producers, large consumers, traders and TSOs. It enables trade from two hours after the day-ahead market closes until one hour before delivery.

The instantaneous changes are handled by the TSO. The balancing services provide the real time balance between production and consumption; this is done through use of the reserves for frequency control. In the Nordic region one distinguish between primary, secondary and tertiary control. The primary reserves are activated automatically with an activation time from a few seconds to a few minutes. They are followed by activation of the secondary reserves, which makes the primary reserves available to handle new imbalances. In Norway the secondary reserves are activated manually, the activation time is from 2-3 minutes up to 15 minutes. The tertiary reserves are also activated manually; they have an activation time from 15 minutes. The reserves are bid into a common Nordic regulating market controlled by the TSOs.

Chapter 8: Ancillary services

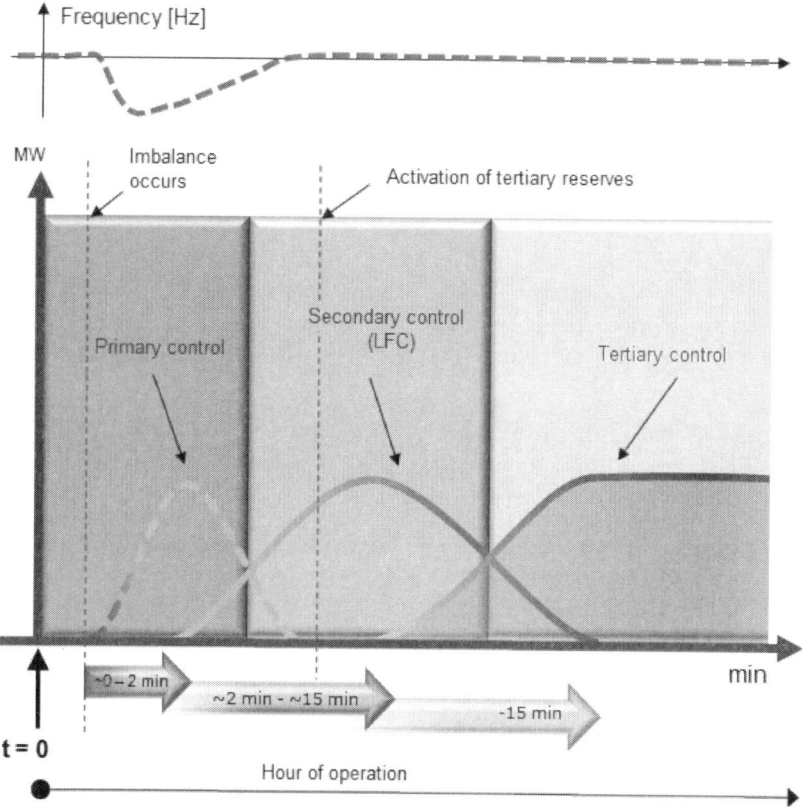

Figure 8.3 Principal overview of the activation of the reserves after an imbalance

The sequence of the activation of the reserves for balancing is shown in Figure 8.3. Assume a contingency, for instance the loss of a generating unit occurs. That leads to a drop in frequency. The red dotted line in the upper part of the figure shows the frequency develops further on. The drop in frequency is detected by the system and leads to immediate activation of the primary control as explained earlier. The primary control will not restore the original frequency completely. There will be a small frequency deviation and a small imbalance left. It is left to the secondary control to re-establish the exact balance and restore the correct frequency. At the same time the primary reserves activated in the first stage, are returning to stand-by mode. So the original amount of primary reserves is re-established. As indicated in Figure 8.3 secondary control can be automatic, based on Load Frequency Control (LFC) and making use of the Area Control Error

(ACE) which is explained earlier[3]. Finally, tertiary control is used. That makes it possible to return the secondary reserves to reserve mode as is also indicated in the figure. The tertiary reserves are assumed to be in use without any time limit

8.6 Costs

8.6.1 Costs for Active Reserves

The cost of providing active reserves, including control equipment, consists of three components:

- Investment costs.
- Operational costs to keep the reserve on stand-by.
- Operational costs when the service is activated.
- Opportunity cost to keep generating unite in stand-by in-stead active production.

The operational costs to keep the reserves on stand-by, include in some situations, costs for personnel and other operating costs to keep the units ready to start within 15 minutes. These costs are also low compared to the total investment.

These three cost components are different in thermal and hydro systems, so we will discuss the two categories separately.

Hydro

Investment Costs

The investment costs consist of the capacity cost, and also the cost of control equipment and other items that are necessary to be able to participate in the service.

Estimates of Norwegian hydro capacity costs (new developments) indicate investment costs from 1000 to 3000 NOK/kW.

The control equipment is a very small part compared to the total capacity costs, and the equipment is usually installed independent of whether you actually want to participate in the Ancillary services provision or not, because some of the equipment is needed in order to start and synchronize the unit. A rough estimate indicates that equipment necessary to deliver active reserves is about 2% of the

---

[3] Automatic secondary control is not yet implemented in the Nordic system. There are plans to introduce it.

total investment. Even if these estimates are inaccurate, we conclude that control equipment is a small part of the total investment.

Operational costs to keep the reserve stand-by
The cost of keeping capacity reserve on stand-by in the production system is influenced by how the unit efficiencies depend on output. In this matter there is a difference between thermal and hydro systems. Figure 8.4 shows efficiency curves of different types of hydro power units. The curves depend on what type of turbine that is used, but common for all is that the maximum efficiency is obtained at a level far below the maximum output. For a Francis-turbine, which is very common in Norway, the best efficiency is at approximately 80% of maximum load. This means that when units are operated at best efficiency, which is wanted in most cases, there is considerable spinning reserve without extra operational costs.

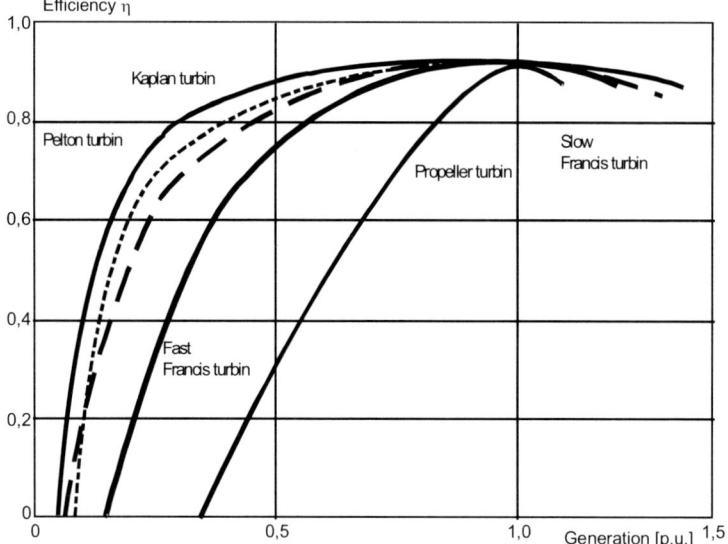

Figure 8.4. Rate of efficiency for different types of hydropower turbines.

The operational costs to keep the reserve stand-by includes in some situations, costs for personnel and other operating costs to keep the units ready to start within a specified time. These costs are also low for a hydro unit.

Operational costs connected to activation of reserves
These costs consist of efficiency related costs generated by deviations from the best efficiency level and also start-up costs related to the need of involving extra units.

The costs caused by the drop in efficiency when reserves are activated, can be high in some cases. We can see from the figure above that the drop in efficiency is small for most types of units. The units with the flattest efficiency curve are also the ones that are usually used for primary control services. However, we must remember that the drop in efficiency affects the whole production from the specific unit, and not only the additional production that is needed to perform the control.

We must include start-up costs when we activate a reserve that is ready to start. These costs consist of abrasion, and a certain amount of wasted water. These costs are small for hydro units.

Thermal Systems

We assume that a thermal system is capacity constrained. This means that capacity kept in reserve must be replaced by a different source to cover the maximum load, the other source usually consists of gas turbines. Even if it is a coal power plant, which is very often used as Primary power control, it would be correct to use a power cost equal to the investment cost for gas turbines.

Major investments are often made in thermal power plants to improve control capabilities, and influence operational costs. This is related to how the control service is conducted.

Primary Control
There are several ways to provide primary control reserves in thermal power plants, but the most common method is to throttle the inlet valves on the high pressure turbine in conventional coal power plants. Extra energy is thereby stored in the boiler in the form of steam pressure, and can be released by fast turning the valves open. The duration of the power increase is limited by the accessible steam volume in the boiler. The negative consequence of this method is that power plants must operate with throttled valves and over-pressure as a standard setting, which reduces the efficiency by 0.5 percentage points, and increases the fuel cost. In addition, extra investments, and increased abrasion as a consequence of the increased production are both additional costs.

Secondary control
The secondary control reserve is normally spinning reserve in a thermal system. It is not normal to include start/stop procedures in the automatic control functions, even if fast starting gas turbines are used as a part of the Secondary control. The cost of keeping a unit spinning or in a hot standby mode is considerable. We must

include considerable start-up costs even for reserve in hot stand-by. These costs are caused by abrasion, and a certain amount of wasted fuel.

Black-start capability
Most thermal generators need to take electricity from the grid in order to start up. Consequently, if there is a total black-out, restart is a problem. The amount of ancillary power needed to start a hydro turbine is much less than what is needed for a thermal turbine, especially a steam turbine. Normally a hydropower plant will be equipped with batteries sufficient to perform a black-start. Therefore, hydro plants can offer low cost black-start capability.

8.7 A simple model for pricing of reserves

8.7.1 Purpose

The purpose of the model described here is to illustrate some basic mechanisms with impact on the price of capacity reserves as well as on other electricity prices. The model is too simple for analysis of a realistic capacity market. The focus is on a system and on a load situation where capacity and capacity reserves are in short supply. Under that kind of circumstances, certain mechanisms will appear which are less important under normal (i.e. capacity surplus) conditions. But it is important to understand these mechanisms. In a more comprehensive and realistic model, these mechanisms have to be included, and the importance will increase with the declining capacity margins we will probably see in the future.

8.7.2 Assumptions and border conditions

Market design

We assume institutional or organizational conditions in line with existing conditions in the Norwegian electricity market:

- There is a Spot Market for electricity.

- There is a market for reserves acquisition (ROM) settled before the Spot Market.

- There is a Balancing Market (BM) where bids are provided after the Spot Market is closed.

- The reserves are activated in merit order based on bids in the BM.

There is a link between these markets caused by the fact that a generating company (and a consumer) can chose what market(s) a given unit should operate in. It can either be in the Spot Market or in the ROM. Once a unit is accepted in the ROM, it is required that it provides bids to the BM.

A simple model for pricing of reserves

There are different ways in which a market for reserves can be operated. One important aspect is the timing. There is presently a one weektime frame for the capacity reservation in the Norwegian ROM.

Another important detail is how to pick the winning bids. In the ROM, that is presently done only on the basis of prices offered for being available (stand-by modus) as reserves. Prices for being activated are not regarded (that prices are not available at that time).[4]

Another important aspect is to what extent a unit accepted as reserve, i.e. accepted in the ROM, is allowed to bid on the Spot Market. This can to some extent be allowed by Statnett today, but there are certain restrictions. The simple model we are describing here is based on the assumption that spot bidding is not allowed for units standing as reserve.

Simplifying assumptions
A set of simplifying assumptions are made. That concerns the costs for the generating units. We disregard costs and restrictions related to start and stop. We assume constant marginal operating cost (fuel cost) from zero to full output. These assumptions are necessary for the simple analysis we are doing here and are used in many similar models, but they represent a serious simplification compared with real life.

Another simplifying assumption is that we disregard the grid. We simply regard a number of generators and consumers connected to the same bus.

We also assume there are a large number of competing units on both sides so there is no market power.

8.7.3 Balance between the Spot Market and the reserves market

The model is based on balance between the Spot Market on one side and the BM and ROM on the other. If there is a balance, there is to be no profit in switching from one side to the other. In the following we consider a single (marginal) unit with a given capacity. Possible profits in the two markets will be (notice that we are here disregarding the possibility of being in the spot market and bid in downward regulation to the BM):

---

[4] If we compare this with normal option pricing, that is equivalent to the settlement of an option price (a premium) without knowing the strike price. The SO will know the price for keeping the reserve available, but not the price for calling it up if it is needed.

Spot Market:

Net income = Capacity • (spot price – marginal cost)

Reserve markets:

Net income = Capacity • Reserve Option price

               –   Operating cost of keeping reserves available for generation

               +   Expected revenue from possible activation (in the Balancing Market)

Net revenue in the two markets must be equal for a marginal unit. Another way to express the same is to say that lost revenue in one market (which represents the *opportunity cost*) is to equal the revenue in the other market. For example: the opportunity cost of keeping a unit as capacity reserve, is the revenue lost because we cannot use that unit to produce and sell in the Spot Market.

Further simplifications

By introducing some simplifying assumptions, it is possible to show how the price of Reserve Options and the price of energy (the spot price) will be, if there is free competition in both markets and it is possible to switch between them. The simplifying assumptions are that we disregard costs for keeping units available for generation at short notice. As mentioned earlier we also disregard start/stop costs and different restrictions affecting the unit commitment. In addition we are not taking possible income from activation of balancing units into account.

With these assumptions we get the following net income in the two markets:

Spot Market:

Net income = Capacity • (spot price – marginal cost)

Reserve markets:

Net income = Capacity • Reserve Option price

Since the income from the two markets must be equal for the marginal unit, it follows that:

Reserve Option price = spot prices – marginal cost

The situation is illustrated in Figure 8.5.

A simple model for pricing of reserves

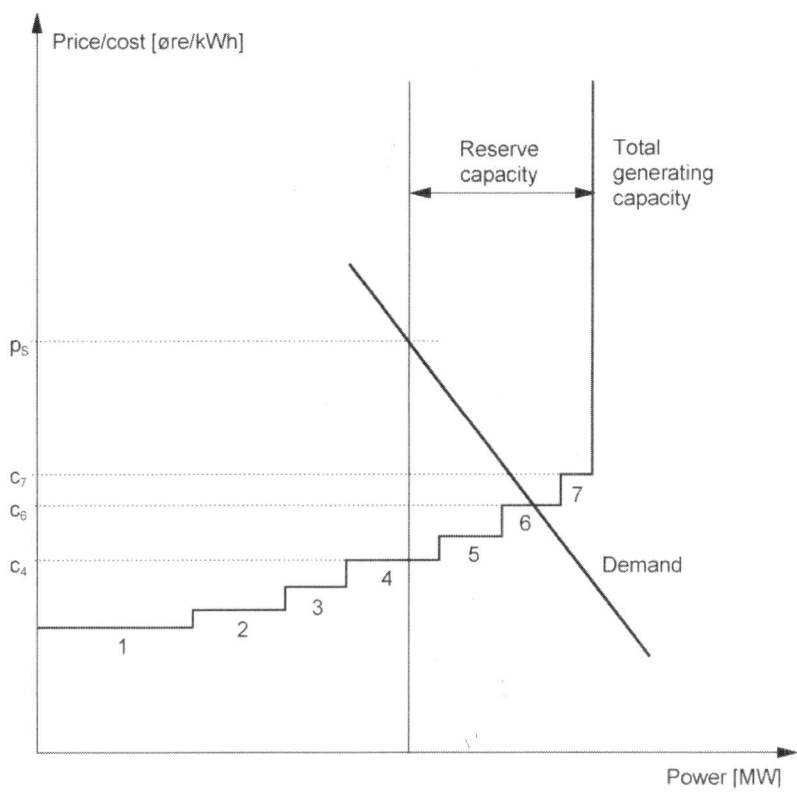

Figure 8.5. Price settlement for energy and capacity reserves.

In the example shown in this figure, there are seven generating units arranged in merit order representing the supply side in the market. Marginal costs for the units are $c_1, c_2, c_3, ..., c_7$. With the demand indicated in the figure, the price would have been c_6 if there had been no requirement for capacity reserve. Unit 6 would have been the balancing unit. With the introduction of a requirement for capacity reserves as indicated in the figure, unit 4 will be the balancing unit. The market price will be p_s and the marginal cost will be c_4. Unit 4 will run on part load and units 5, 6 and 7 will serve as capacity reserve[5]. This is an economically optimal

---

[5] It should be emphasized that this solution is based on strongly simplified assumptions. In practice part of the capacity reserve will be based on several units in operation, which means part load operation. For instance all primary reserve (frequency control reserve) is tied to units in part load operation.

solution. The units with lowest operation (fuel) cost are in active operation while the units with higher operation cost are used as stand-by.

The question is then: what is the corresponding reserve capacity price (ROM price), i.e. what is the price of capacity reserves that will lead to this solution.

It is obvious that the energy price (spot price) will be p_s. In order to have sufficient capacity for active generation as well as capacity reserves, it is necessary to hold back consumption by this high price. This energy price can be regarded as the sum of the marginal cost for the balancing unit, c_4, and capacity price p_c. That means the capacity price will be: $p_c = p_s - c_4$. With this capacity price, unit 4 will be indifferent between being in active production and being standby. In the first case the income will be p_s and the cost will be c_4, giving a revenue of $p_s - c_4$ which equals p_c. In the second case the unit will receive a fixed compensation of p_c. So the two alternatives are equally profitable for that unit.

With this capacity price, all the other units will find it most profitable to stay either in active production (as units 1, 2, and 3) or in standby mode (as units 5, 6 and 7). The profit for units 1, 2 and 3 will be higher if they produce and sell at a price p_s instead of being in standby mode and receive a compensation p_c for that. If units 5, 6 and 7 receive compensation ($p_c = p_s - c_4$) for keeping capacity as reserve, it will be less profitable to enter into active production. This indicates that the capacity price $p_c = p_s - c_4$ represents a stable equilibrium. This reserve capacity price is compensation for being stand-by, but can also be regarded as a compensation for staying out of active production.

European and Nordic co-operation.
There is a long history of co-operation between electric utilities in Europe. The Union for Coordination of Production and Transmission of Electricity (UCPTE) was established as early as 1951. Be-Ne-Lux, Germany, France, Switzerland, Austria and Italy were taking part from the start. The number of countries participating has increased in line with the development of EU participation. In 1999 the name was changed to UCTE reflecting the fact that production was no longer included in the co-operation. Producers were supposed to compete, not co-operate. In 2009 it was included in the European Network of Transmission System Operators for Electricity (ENTSO-E). NORDEL was established in 1961. That was also an organization for coordination of production as well as transmission. Power producers were taken out of the cooperation during the 1990s in line with what happened in the rest of Europe making NORDEL an institution for co-operation between Nordic TSOs. NORDEL was also included in ENTSO-E, which from July 1, 2009 is the organization responsible for co-operation among European Transmission System Operators.

Figure 8.6. Co-operation between electricity utilities in Europe

8.8 The Norwegian and Nordic system

There has been an extensive cooperation on ancillary services between the Nordic countries organized by NORDEL.

The quality requirement on frequency: +/- 0.1 Hz during normal operation and maximum 0.5 Hz frequency drop during contingency. Maximum time deviation 30 seconds.

8.8.1 Primary reserve

Table 8.4 shows the requirement for primary reserve in the Nordic system; the total requirement and the division between the countries. Notice that the eastern part of Denmark is included in this case, not the western part which is synchronous with the Continent. We distinguish between two classes of reserves: Normal reserves and contingency reserves. The normal reserves are supposed to keep the frequency deviation below 0.1 Hz, which means the frequency should be between 49.9 and 50.1 Hz.

Table 8.4. Primary reserves in the Nordic system

	Frequency bias (Load/frequency characteristic) (MW/Hz)	Normal frequency control reserve (MW)	Contingency reserve (MW)
Denmark (east)	250	25	179
Finland	1250	125	205
Norway	2000	200	313
Sweden	2500	250	303
NORDEL	6000	600	1000

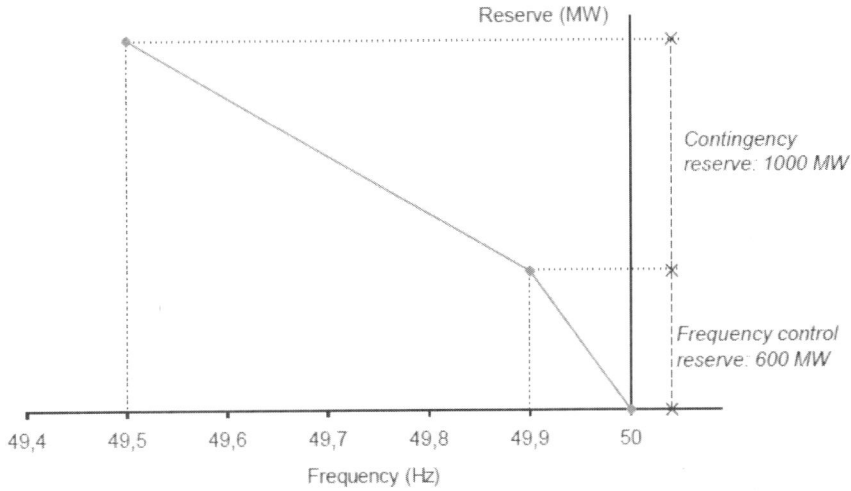

Figure 8.7. Frequency response in the Nordic system

8.9 Some experiences

The System Operator, Statnett, has from an early stage of the market liberalization been responsible for the Balancing Market (BM) in Norway. In November 2001 Statnett established a new power reserves market, the Reserve Option Market (ROM).

In this section we describe practical arrangements and experiences concerning these two markets. There have been some changes over the years and the following description is based on the rules applied before August 2004.

8.9.1 The Balancing Market (BM)

The BM is used for up and down adjustments in real time and is based on bids from generators and consumers.

Bids to the BM for the subsequent night and day (from 24.00 to 24.00) are to be submitted before 19.30 in the evening. By that time the Spot price is known. It is required that the price for up-regulation (or for reduction of consumption) is higher than the spot price and the price for down-regulation (or increase of consumption) is lower than the spot price. There is an upper limit of 50 000 NOK/MWh (50 NOK/kWh)[6]

This upper price limit is very high, and far above any estimates of VLL. That means that VLL will represent the practical limit. There is no reason for a rational consumer to buy balancing power at a price higher than the cost imposed on him by load shedding. The problem is of course that in many cases the consumer will not see that price.

The minimum quantity is 25 MW. The minimum duration is one hour. The bidder must be able to provide the offered quantity within 15 minutes.

When there is a need for activation of balancing power, the bidders are called up, normally by phone. The available bids are normally activated in merit order. The highest activated bid determines the balancing price.

---

[6] Even that limit can be exceeded if the Spot price is extremely high. The limit is then two times the Spot price (if that is more than 50 000 NOK/MWh). In principle there is no upper limit to the Spot price. For certain technical procedures Nord Pool has a price limit of 16 500 NOK/MWh (2000 €/MWh), but that can be exceeded if necessary.

Some experiences

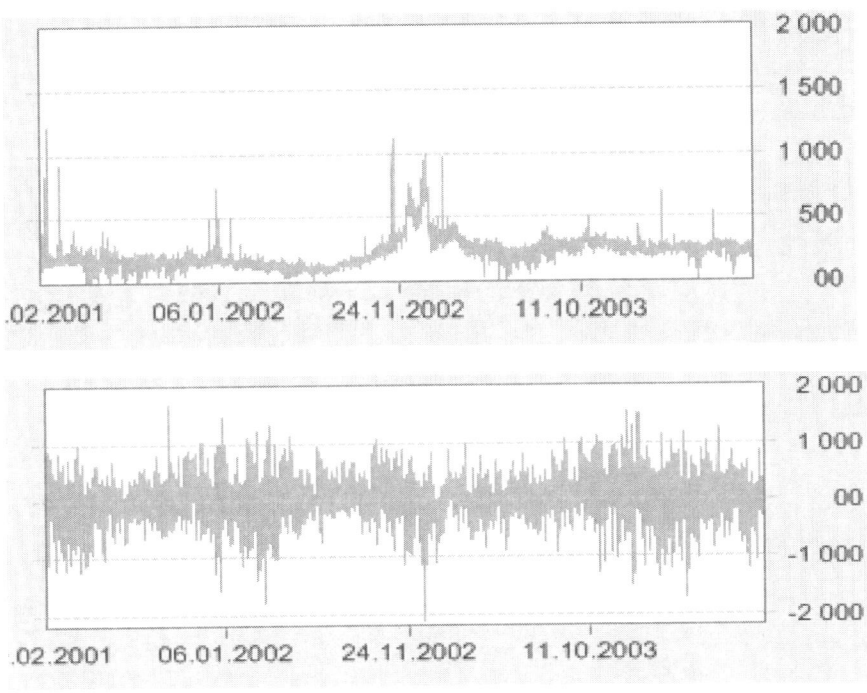

Figure 8.8. The Norwegian Balancing Market from February 2001 to August 2004. Prices in NOK/ MWh (upper) and quantities in MWh (lower). Source: Statnett

8.9.2 The Reserve Option Market (ROM)

This description is to a large extent based on [5]. In addition the development over the last few years is described.

As long as there was a safe capacity margin in the Norwegian and Nordic system, there was always ample supply of Balancing Power. The bids could cover the needs at any time.

As the balance became narrower, Statnett could see indications that the provision of reserves on a voluntary basis (or more precisely: on the basis of incentives from the uncertain Balancing Market) would be insufficient. It was therefore decided that action should be taken to make sure that sufficient balancing reserves should be available at any time.

Alternative solutions

Alternative solutions were considered. Initially agreements were made with selected generators to make such reserves available during specific predefined periods in return for certain remuneration. The market participants opposed this arrangement. There was a lack of flexibility. The reservation was to some extent unpredictable. Purchasing reserves on the spot market was also considered. A model with an "uplift" in the spot price was rejected for principle as well as practical reasons. It is more complex and time consuming to change Elspot, which is a Nordic market, than to find an independent national solution.

Long-term contracts were also considered, preferably contracts giving the generating companies incentives to invest. One other possibility was to make investments in gas turbines.

The focus turned, however, more to disconnection of industrial consumption. Agreements on reduced tariffs for interruptible consumption have been used for a long time. These apply in particular to electric boilers that can be disconnected at 2 or 12 hours notice. The possibility for including some industrial consumption in the BM would require shorter response time. In general, the industry showed a keen interest in participating and finding a solution to the power reserve problem. Initially some individual agreements with a few major companies were made.

Statnett then found that the most appropriate solution was to introduce an arrangement where suppliers of reserves, from the generating as well as the consuming side, could compete on supplying the system with reserves. That was the origin of the Reserve Option Market (ROM)

Development of the ROM

Statnett started with two tendering rounds for power reserves, in both cases with different time horizons. One type had three months duration and the other one had one year. The contracts from these two rounds were effective from 1 November 2000 and 1 January 2001 respectively.

Consumers and generators were asked to offer reserves for the regulating market (i.e. what was defined as Secondary reserves at that time, but is now defined as Tertiary reserves). The bids did not have to be linked to specific nodes, but to one of three grid regions, A; B or C[7]

In addition the following conditions were used:

- The power reserves should be made available from 6 a.m. to 10 p.m. Monday through Friday

---

[7] The subsequent bids on the Balancing Market have to be tied to one specific node.

- Minimum volume: 25 MW
- Further requirement on the reserves:
  - Must be possible to activate within 15 minutes.
  - Full output for at least 1 hour without interruptions
  - Full output for at least 10 hours pr. week

ROM included only reserves for up-regulation (sufficient reserves for down-regulation were assumed to be available without special provisions), which means that generators must have available capacity to increase generation and consumers must have consumption that can be reduced. On the consuming side, however, it was possible to reduce the normal consumption level due to for instance high Spot price, and thereby reduce the available reserve. Statnett should in those events be notified as quickly as possible and the remuneration should be reduced corresponding to the duration of the reduced quantity.

If an offer for reserve was accepted, the supplier had an obligation to provide bids to the Balancing Market. But there was no requirement that the supplier should indicate (at the ROM stage) a price level for these subsequent BM bids.

Principle for selection of bids

Statnett is responsible for determining the need for Ancillary services in general and that includes Tertiary reserves, which is discussed here. Statnett determines the need for this type of reserves[8] and makes an assessment on whether pricing is needed in order to acquire the needed reserves. There will always be some uncertainty affecting this assessment and that increases with increasing time horizon.

After having received the offers from both sides (supply and demand), Statnett makes an initial merit order list based on prices alone. If the outcome of that gives a too dominant share from one of the sides, this initial list must be adjusted. Statnett requires that each category has a minimum share of the ROM, and if one is too dominant, some of the initially accepted offers from that side will be replaced with bids from the other.

Due to possible congestions in the transmission grid, there are requirements concerning the geographical distribution of reserves. Norway is divided into three regions, A, B and C (the regions are denoted by letters in order to avoid confusion with the Elspot-regions) and Statnett defines reserve requirement for each region. Certain technical factors can also influence the selection of bids.

---

[8] The need for reserves is generally based on NORDEL recommendations.

All the accepted bids within one region are awarded at the same price. There is a difference between regions, but no difference between competing bids within regions. In particular we notice that bidders from the generating and consuming sides receive the same price.

After this price clearing process, the results are published on Statnett's website.

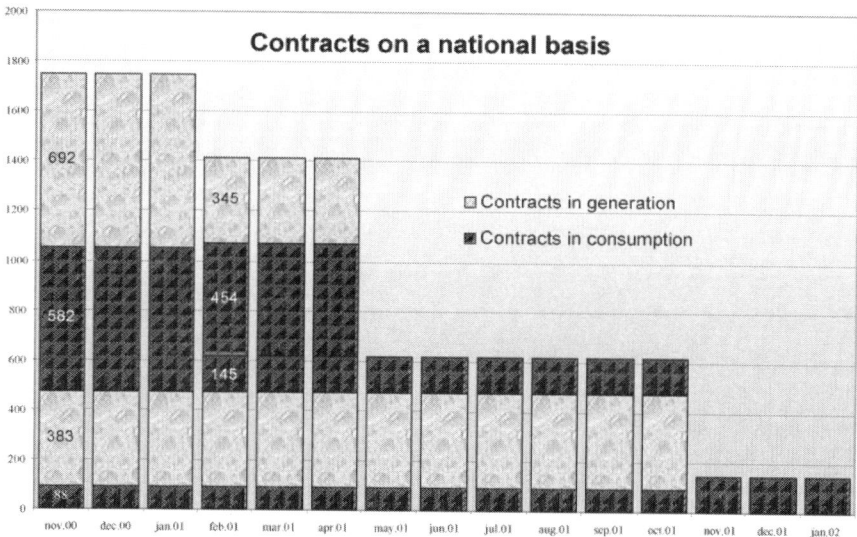

Figure 8.9. Contracted amount of reserves from November 2000.

8.9.3 Special conditions for consumption side reserves

Consumers offering balancing power are to be fully activated within 15 minutes from when the order is given, exactly as the generators, see [7]. But when the activation period is over, it is left to the consuming entity itself to decide how quickly normal consumption will be restored. The payment according to the current BM price will terminate 15 minutes after order is given to deactivate.

Some experiences

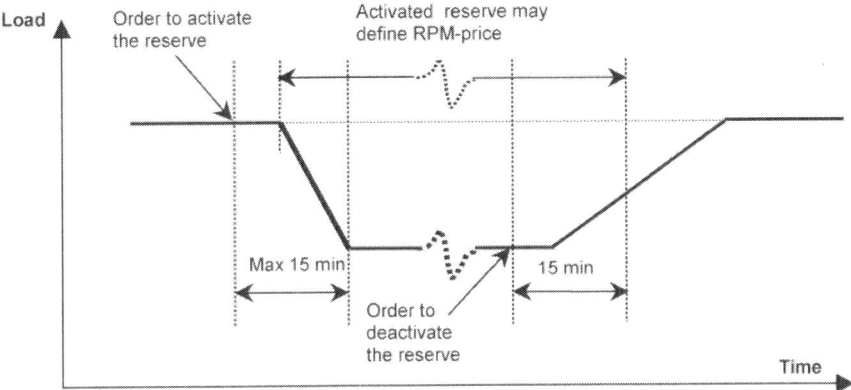

Figure 8.10. Activation of consumer side reserves.

Consumer side reserves can claim a rest period after activation of maximum 8 hours. Under the new market arrangement from October 2004, this possibility is made price dependent and included in a more general set of rules.

8.9.4 Results from the first two rounds

In the first tendering round where offers were requested for the months November 2000 to January 2001, 110 companies were invited. That resulted in contracts for totally 1745 MW for that period (including 471 MW one year contracts). The second round comprising February to April 2001, led to contracts with 11 suppliers for a total of 944 MW. In this second round 23 out of 80 offers were accepted. In general the market responded well to these two first rounds for tender. Far more capacity was offered than Statnett needed. Both sides – generation and consumption – were participating. The selection of suppliers was based solely on price, which means there was a good distribution geographically and between the two sides.

Chapter 8: Ancillary services

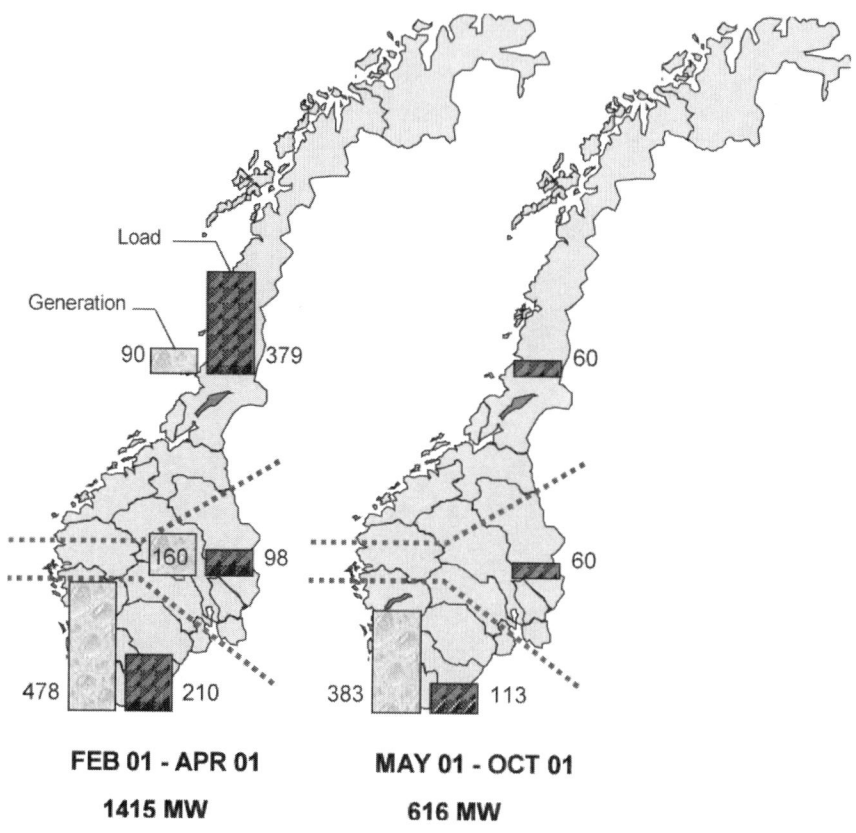

Figure 8.11. Reserves from load and generation in different areas.

Statnett was generally satisfied with the outcome of this initial period. The volume offered and the number of companies involved, indicated that there was considerable competition in supply of reserves. All companies involved followed up with daily offers to the BM and temporary unavailability has been reported according to the contract terms.

During this season, or more specifically on Monday 5 February 2001, the highest demand ever experienced in the Nordic system, was recorded. The Norwegian as well as the Nordic system had a satisfactory power balance, included sufficient reserves, but it was close to the limit.

Some experiences

8.9.5 Development during the next few years

After the winter season 2000/2001 the ROM market has been in operation on a regular basis. The time resolution has been 1 month, 3 months and 12 months. The price for 1 month options during a period of 3 years is shown in Figure 8.12.

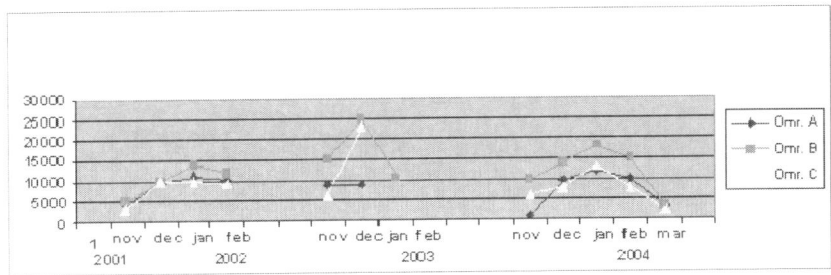

Figure 8.12. Development of Regulating Option Prices (NOK/MW month) from November 2001 till March 2004. The market is only in operation in the winter, normally from November till March.

There is a slightly increasing trend, but the time span is too short to draw a definite conclusion.

There are some interesting details shown in Figure 8.12. We notice that ROM price is normally highest in January, which is to be expected because the yearly peak load is expected to be in that month.

Another interesting observation is the development in the winter 2002/2003 which was a period with high spot price due to low precipitation in the autumn of 2002. We notice that the ROM price was high in December, but after New Year, the price went down or the market was not active. (Where there is no indication of price in the figure, the market is inactive, which means that Statnett finds it unnecessary to call for ROM bids. Sufficient reserves are expected without this price incentive.) The explanation is that due to the high spot price, the consumption and therefore the peak load went down. The capacity margin was high and sufficient balancing reserves were expected to be available.

During the whole period, generation as well as consumption has contributed to the ROM, but the consuming side has been largest, in certain periods with up to 75 % market share. Another striking difference between the two sides is that bids into the BM tend to be lowest on the generating side. The tendency is that consumers offer a low price no ROM and a high price on BM and vice versa on the generating side. This can lead us to suspect that the consuming side is not really interested in taking part in the regulation. They bid so high into the Balancing Market that they can be confident they will not be called up to contribute.

However, Statnett has a contractual right to make a test of their preparedness to contribute; once a year, consumption can be ordered down on instruction from Statnett.

8.10 References

[1] I. Wangensteen, O. Wolfgang, G. Doorman: "Capacity pricing in a free market" TR A60137, January 2005.

[2] G. Doorman: "Peaking Capacity in Restructured Power Systems", Dr.ing. thesis, NTNU, November 2000

[3] E. S. Amundsen, Lars Bergman: "Metoder at säkra toppetfekt på avreglerade elmarknader" Rapport september 2001

[4] C. Pass, B. Lowes, L. Davies: "Dictionary of Economics" Collins 1988.

[5] G. Nilsen, B. Walther: "Market-based Power Reserves Acquirement" Conference on: Methods to Secure Peak Load Capacity in Deregulated Electricity Markets. Saltsjøbaden, Stockholm June, 2001

[6] N. Flatabø, G. Doorman, O. S. Grande, H. Randen, I. Wangensteen: "Experience with the Nord Pool Design and Implementation", IEEE Transactions on Power Systems, Vol. 18, No. 2, May 2003.

[7] G. L. Doorman, O. S. Grande, I. Vognild: "Market based solutions to maintain system reliability in the Norwegian power system", CIGRE 2002, Paris, 25-30 August 2002.

[8] I. Wangensteen: "Markedsbaserte løsninger på effektprising", TRA5658, SINTEF Energy Research. 2002.

[9] Hal R. Varian, *"Microeconomic Analysis"*, (3rd ed), New York: W.W. Norton & Company,

[10] S. Stoft: *"Power System Economics"*, IEEE Press, 2002.

[11] S. Stoft: *"The Demand for Operating Reserves: Key to price Spikes and Investments"*, IEEE Transactions on PAS, vol. 18, No.2, May 2003.

[12] Ove S. Grande, Gerard Doorman: "Reduserbart forbruk som reguleringsobjekt. Sammenhengen mellom RKOM, Elspot og RK", SINTEF Energiforskning AN 12.03.64

[13] Laurens J. de Vries: "Securing the public interest in electricity generation markets", PhD thesis, Technical University of Delft, June 2004.

References

[14] I. Wangensteen, O. Wolfgang, G. Doorman: "Capacity pricing in a free market", TRA6037, SINTEF Energy Research, December 2004

[15] I. Wangensteen, O. S. Grande: "Provision and Pricing of Ancillary Services in a Deregulated Hydro dominated System" CIGRE symposium on Impact *of Open Trading on Power* Systems,Tours, June 1997.

[16] B. Bakken: "Technical and economic aspects of operation of hydro- and thermal power systems" PhD-thesis, Norwegian University of Science and Technology, December 1996.

[17] Gjerde, Fismen, Sletten: "System operator responsibility in a deregulated power market" CIGRE symposium on Impact of Open Trading on Power Systems, Tours, June 1997.

9 Costs and regulation of grid monopolies

9.1 Grid costs

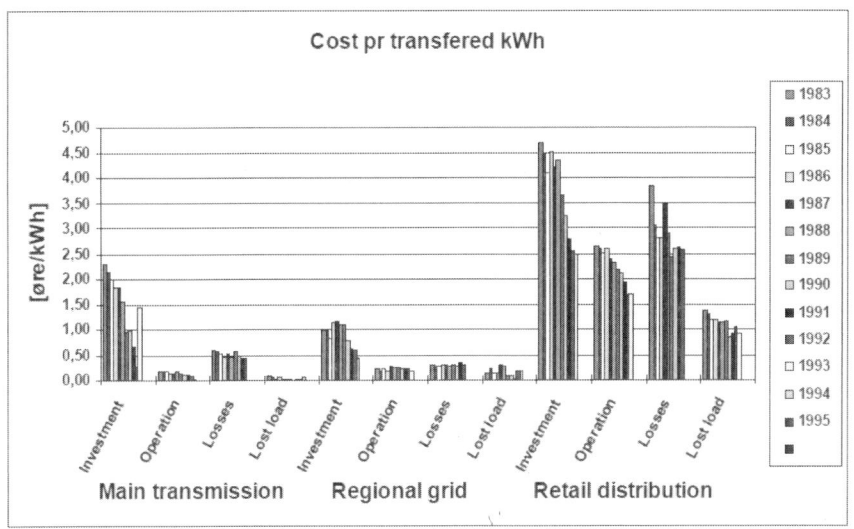

Figure 9.1. Grid costs in Norway. Overview.

Figure 9.1 gives an overview of the Norwegian grid costs for a period of 12 years. The costs for all years are adjusted to the 1995 price level.

The costs are divided into 1) investments, which means actual investments each year, 2) operation costs (maintenance, repair etc.), 3) transmission losses and 4) losses for the consumers due to interruptions. It is also distinguished between the transmission grid (mainly 420 – 220 kV), the regional grid (132 – 60 kV) and the distribution grid, which comprises high voltage (10 – 20 kV) and low voltage (230 V).

It is evident that retail distribution is by far the most costly part of the grid and dominates for all cost categories.

This pattern is typical for other countries too. In countries with a more concentrated population, the costs for the main grid can be lower. Distribution grids in urban areas are often based on underground cables. In that case the investment costs are high but the operation costs and interruption costs are low.

The grid costs can therefore be somewhat different from this Norwegian example, but the important characteristics are the same.

9.2 Economy of scale in electricity distribution and transmission

It is generally accepted that the distribution grid is a natural monopoly. That means it is not economical to build parallel distribution grids in order to create competition between grid companies.

The fact that the distribution grid is a natural monopoly does not mean there should be only one distribution company in the country. It is a local monopoly in the sense that there should be only one company covering one area. Figure 9.2 gives an illustration of the scale economy for a distribution utility serving one specific area. It is taken from a design study [5] where a distribution grid is optimized for a set of alternative demand scenarios. The figures indicate that there is only a small increase in total costs for a substantial increase in the amount of distributed power. That is the case for an increase in peak load and the same happenes with an increase in distributed energy per year.

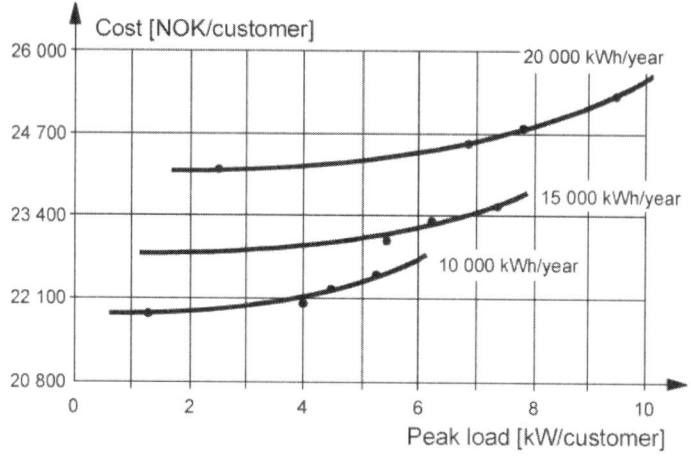

Figure 9.2. The cost of an optimized distribution grid for alternative load scenarios [5].

We can also define the scale economy for a distribution company in relation to how the cost depends on the size of the company. How large an area should a utility cover? This has been investigated and *Figure 9.3.* shows the result of an

Economy of scale in electricity distribution and transmission.

empirical study for Norwegian utilities [2]. According to this study there is a cost advantage of increasing the increasing the size to about 10 000 customers.

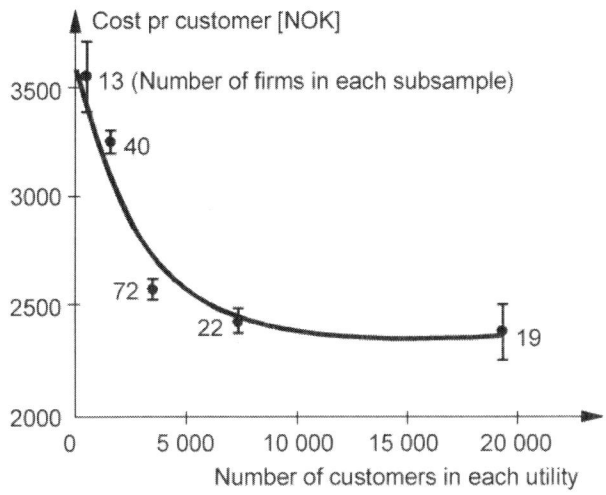

Figure 9.3. Distribution cost for utilities as a function of an increasing number of customers. Empirical study for Norwegian distribution utilities from 1988 [2].

From a scale economy point of view, there is no reason to go for a large scale horizontal integration of distribution utilities.[1]

The scale economy for the transmission grid is not necessarily comparable with the distribution grid. Up to a certain limit, one can increase the voltage level and one can increase the cross-section of lines. The capacity of a transmission line will then increase more than proportionally with the cost. However, once the limit is reached one has to build parallel lines. That is typical for the Swedish grid where we find a set of parallel lines from the hydropower sources in the north to population centres in the south. The scale economy of the transmission grid is therefore questionable. But for different other reasons it is a rational solution to place the responsibility for the operation and planning of the main transmission grid on one single company. In practice the transmission companies are therefore treated as monopolies.[2]

---

[1] The empirical investigation referred to here is from Norwegian utilities and dates back to the 1980s. Things have probably changed since then and the limit of 10 000 customers is probably not appropriate today. But other investigations support this finding that there is an upper limit on economical size for a distribution utility.

[2] There is some discussion of opening up the transmission grid for so-called Independent Transmission Providers (ITPs). This is a complex issue, especially if it is be implemented

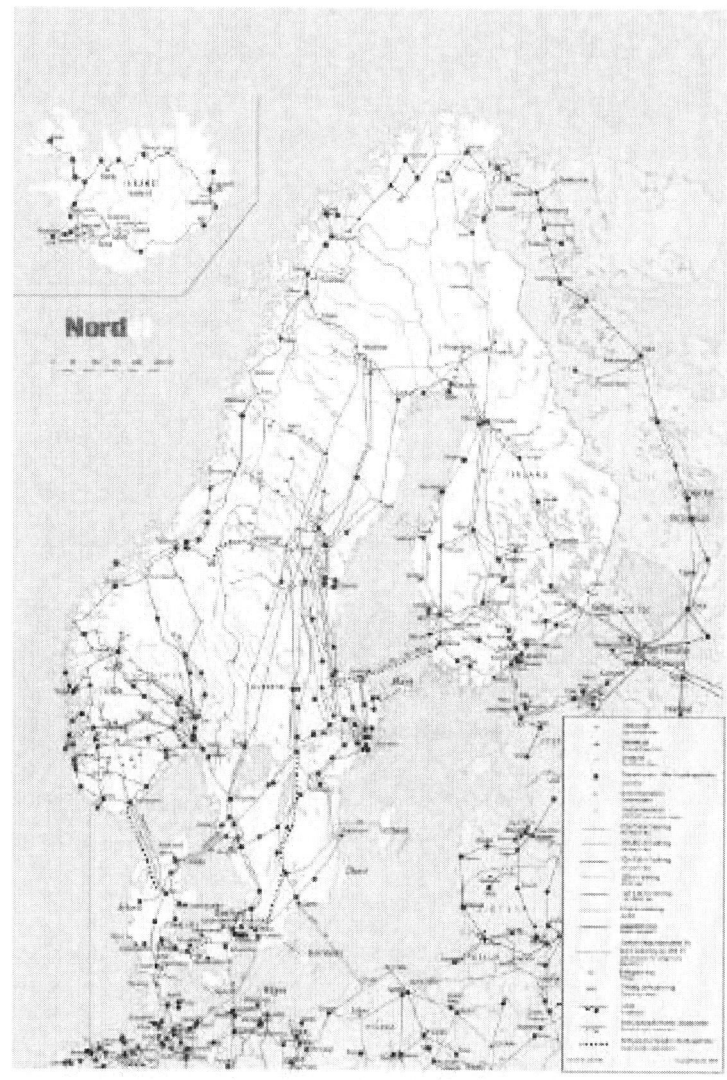

Figure 9.4. The Nordic transmission grid [Nordel]

Deregulation in the Nordic region (as in other parts of the World) included an unbundling of the system. The grid activity was separated from generation and

in a meshed transmission grid. In any case one single entity will be responsible for system operations.

organized as regulated monopolies; one national monopoly for the transmission grid and a number of regional/local monopolies for the distribution. Economic regulation was an important element. In Norway, the Norwegian Water Resources and Energy Directorate (NVE) was given responsibility for that.

9.3 Monopoly regulation

9.3.1 Definition

In economic terms, a natural monopoly is characterized by a decrease in Average Total Cost (ATC) with increasing quantity. That means that Marginal Cost (MC) is lower than ATC over the interval we consider, see Figure 9.5.

In a case like this, it is more costly to divide the production of a given quantity between two or more companies than to produce everything in one single firm. From a cost point of view it is more attractive to operate with a monopoly than with a number of competing firms.

Monopoly regulation in this context means regulation of natural monopolies.

9.3.2 Pricing for a natural monopoly

As mentioned above, the Marginal Cost in a natural monopoly is lower than the Average Cost. This creates some problems related to pricing, see Figure 9.5..

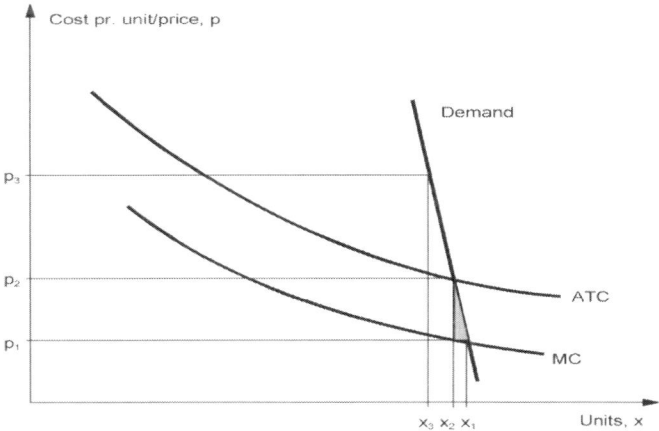

Figure 9.5. Pricing for a natural monopoly.

The figure shows demand, Marginal Cost, MC and Average Total Cost, ATC. If the price equals MC, (p_1 in the figure) we will reach the optimum as described in Chapter 2: maximum social surplus. The problem with this solution is that total cost is not covered. There will be a deficit represented by an area between p_1 and p_2. (~ $(p_2-p_1)x_2$) Normally a company cannot live with a loss like that.

In order to cover the cost, the price can be raised to p_2 which will reduce the quantity to x_2. In that case the cost is covered because the price equals ATC, but there is a loss in social surplus represented by the shaded area in the figure. The figure also shows that a steep demand curve (small price elasticity) contributes to a low social loss.

There are different solutions to this problem. One possibility is to let the community (state, municipality) cover the deficit which is probably an unrealistic solution in most cases. Another possibility which can lead to an almost optimal (second best) solution is so-called Ramsey pricing which was described in Chapter 5. That requires price discrimination which can be difficult to accomplish.[3] It is also possible to use a tariff with different elements. A high fixed element and a correspondingly low variable element is favourable in this respect.

But it seems to be generally accepted that it is not realistic to reach an optimal solution implying marginal cost pricing in the distribution grid. With the low price elasticity we have in electricity demand, the social losses are not unacceptable if prices equal ATC instead of MC.

9.3.3 The objective for monopoly regulation

We can broadly distinguish between two mechanisms causing economic inefficiency for a monopolistic company:

Firstly, the price can be too high. The price is not established in a competitive market. A monopolist has the possibility to overprice his/her product and can thereby cause economic losses compared to an efficient market as we have seen in Chapter 2. This is called *market inefficiency*.

Secondly, the cost can be higher than necessary. As long as any cost can be transferred on to the consumers, there is no strong motive to reduce costs.[4] This type of inefficiency is called *X-inefficiency*.

---

[3] A rude type of Ramsey pricing is being used in Norway in form for so-called curtailable power. That is typically used for supply to boilers with electricity and oil as alternative fuels. Oil is used when electricity prices are high. That type of electricity consumption is sensitive to price changes. A special grid tariff with a price close to the marginal cost is used for this type of consumption.

[4] It should be emphasized that these mechanisms are based on the assumption that the companies are profit seeking. That was not originally the objective for public utility

X-inefficiency can again be caused by different factors:

- Wrong scale. The firm can be too small (in principle also too large) compared with the optimum size. See Figure 9.3.
- Technical inefficiency i.e. use of larger quantities of production factors than necessary.
- Cost-inefficiency, i.e. uneconomical composition of production factors.

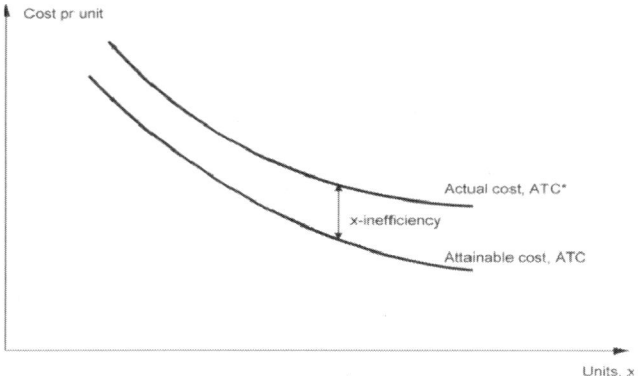

Figure 9.6. X-inefficiency

Figure 9.6 shows how ATC can be affected by X-inefficiency.

The consequences of market inefficiency and X-inefficiency are shown in Figure 9.7. Market inefficiency can appear in form of "overpricing" (p_3 in the figure) and the corresponding loss to society is indicated by the triangular shaded area. We then compare this with price p_2 which can be regarded as the correct one here. As said before, with a price insensitive demand, this loss is small. On the other hand, there is a large transfer of money from the consumers to the utility. This does not bring a societal loss, but it can have unwanted distributional implications.

---

companies. They had an economic objective based on cost minimization. This has in most cases changed after deregulation.

Chapter 9: Costs and regulation of grid monopolies

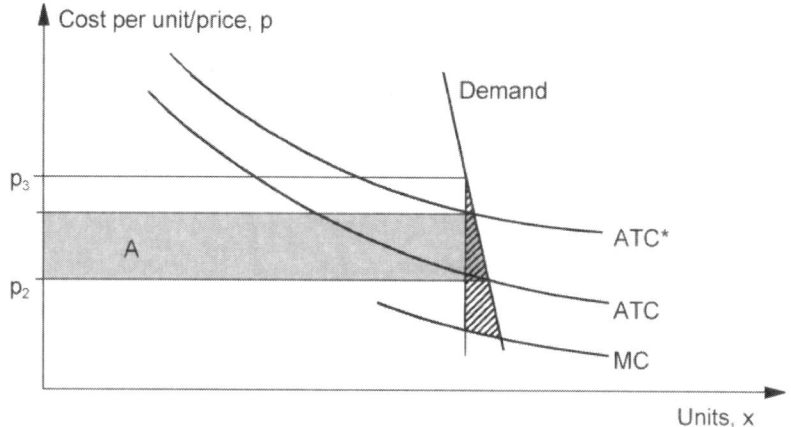

Figure 9.7. Consequences of market inefficiency and X-inefficiency

The losses caused by X-inefficiency are shown by the rectangular shaded area. What should be noticed is that all costs that can be saved represent a loss to society. It is a waste of resources. One important aim is therefore to provide incentives so these unnecessary costs are reduced or avoided.

The main objective of regulation is to eliminate market inefficiency as far as possible, mainly due to its distributional implications, and X-inefficiency mainly in order to avoid a waste of resources.

As mentioned, it can be argued that the correct price is p_2. This covers the exactly cost for an efficient utility. It can also give the company good incentives to reduce costs, but this depends more on how that price is fixed than on the correct level. As we can see from Figure 9.7, it can be more important to give a good incentive than to hit exactly the right price level. The economic benefits from cost improvements (which depend on incentives) are more substantial than benefits from better market efficiency. If we hit the wrong price level, it can have distributional implications, but with respect to economic efficiency it is not so important.

9.3.4 Presumptions

In international literature, it is often taken as an implicit assumption that grid companies are profit-seeking entities. That can be a reasonable assumption for utilities owned by private investors. However, in Norway, and to a large extent in other Nordic and European countries, the distribution utilities are usually owned by municipalities and other public entities.

Monopoly regulation

The purpose of public regulation of a publicly owned monopoly can be discussed. Before restructuring, it was a common objective for utilities to supply consumers with electricity at lowest possible price. Overpricing in the sense that prices were higher than costs was never a problem. The general impression was that low prices, which again required low costs, were given priority.

But still there was an impression that further cost savings could be made if the grid companies were given the right incentives. The *revenue cap* regulation that was introduced in Norway from 1997 was based on that assumption. At the outset, most companies did not change their objective (minimizing cost for consumers), but it appeared that under pressure from a tight revenue cap, considerable costs savings were still made. On the other hand, some utilities deliberately earned less than the revenue cap, which is a clear evidence that their objectives were not profit maximization.

The assumption that public utilities are profit-seeking entities is therefore a simplification. Some have made an explicit change of objectives and introduced profit maximization. Some have not made it explicit, but act as if they have. Others remain old-fashioned cost-minimizing utilities. Still, monopoly regulation is based on the assumption that grid companies are profit seeking entities. It is correct for many of them and incentives that are created by regulation based on that assumption, have positive impact in any case.

In the preceding sections we stick to that assumption, but we should keep in mind that reality is more complex.

9.3.5 Alternative mechanisms

Introduction

There are different mechanisms for regulation. With respect to incentives for cost reduction, which is important in this context, the alternatives have different qualities. Some mechanisms bring weak incentives or no incentives at all. *Cost-of-Service (COS)* regulation is one example. So-called *Cost-plus-regulation* and *Rate-of-Return (ROR)* are basically in this category. They are based on the principle that the price or revenue is tied to the actual cost. Therefore, any cost savings will eventually lead to lower prices/revenues. Incentives can even be negative.

In order to create the right incentives, the regulation should not be tied directly to what the cost is, but rather to what the cost should be. There are different mechanisms. One is use of performance indicators which can be found by benchmarking.

Regulatory lags and *sliding scale regulation* also include mechanisms that, at least to some extent, provide the right incentives. But there are certain draw-backs too.

The regulation is in many cases directed at the price, in some cases at the revenue. Applying the terminology from control theory, we can call these variables *control variables*. If the price is chosen as the control variable, in the short run the company will have an incentive to increase the volume. If the revenue is used, this incentive is no longer the same.

The revenue regulation system presently used in Norway combines different mechanisms that will be described later. First, a discussion of the principles.

Cost-of-Service regulation

We use Cost-of-Service as a general term covering Cost-plus and Rate-of-Return regulation. It implies that the company will have its full cost covered, plus in many cases a small addition. The principle has been used in many countries. The point is that in order to draw capital to the sector, investors must be given a fair return on investments. In an open economy the grid companies must compete with other sectors for capital. The return must be at least as good as for alternative use of capital.

The return on investment, which is subject to regulation, is therefore tied to the market price on capital. In addition the risk can be taken into account. The discount rate on Government bonds can be regarded as risk free. In Norway, the rate for Government bonds plus a risk premium of 2% has been used.

It is obvious that Cost-of-Service regulation provides no incentive to reduce X-inefficiency. In order to attract capital, it can be tempting for the regulator to allow high return on investment. Underinvestment can be serious. If the return is higher that alternative use of capital, it is attractive for the regulated firm to systematically over invest. It can lead to what is some times called "gold plating".

Regulatory lags

With a Cost-of-Service regulation, there is normally a delay from the time the cost arises until the regulation of the price or revenue is effective. Costs have to be taken from the company accounts, which is ready sometime into the year after the cost occurs and there is still some delay until the regulator takes action. In case a company has accomplished a cost reduction in year n, the company keep the extra profit that year and probably also year $n+1$ and a lower income or price level is introduced in year $n+2$, see Figure 9.8.. The regulatory lag makes it possible to benefit from cost savings during two years and that creates a certain cost reduction incentive.

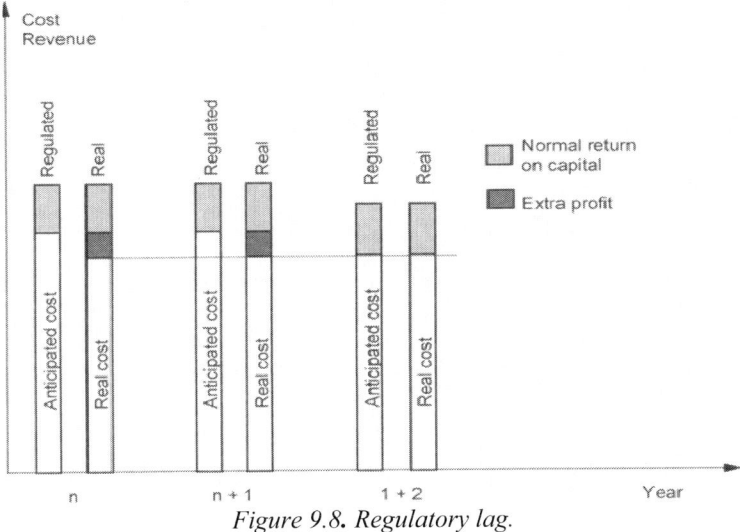

Figure 9.8. Regulatory lag.

Sliding Scale

Sliding scale is a mechanism that divides the benefit from cost savings between the utility and its customers. If we start with the regulatory lag example (Figure 9.8), we observe that after the lag, all the benefit from the cost savings goes to the customers. The sliding scale mechanism divides the benefit between the two parties also for the years after the lag, see Figure 9.9. Thereby the incentive for cost savings is strengthened.

Chapter 9: Costs and regulation of grid monopolies

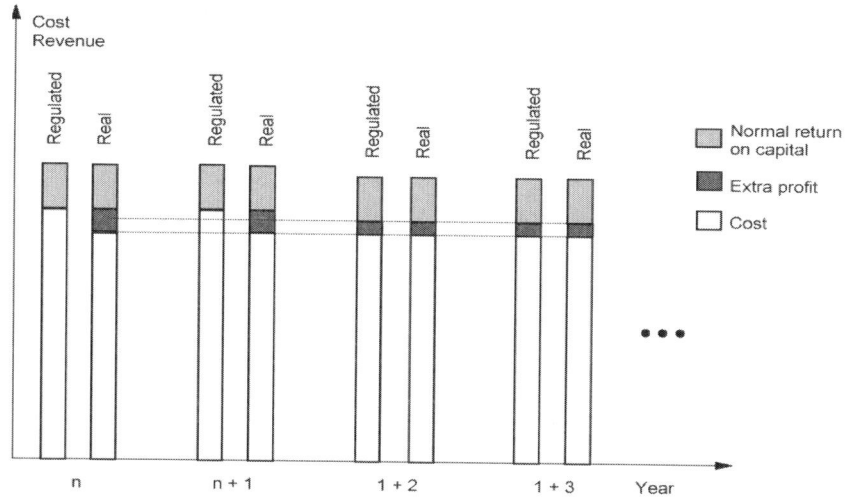

Figure 9.9. Sliding scale.

We define a constant k, $(0 < k < 1)$, representing the share of a cost saving that goes to the customers. The decision on the value of k must be based on a compromise between a wish to give strong incentives to the utility, (k is low) and a wish to give the customers a large portion of the cost savings (k is high).

Regulation based on performance

The idea for this type of regulation is that the revenue or price should be based on how well the company performs. This performance standard or reference should be found outside the company, which requires the use of a benchmarking mechanism.

Benchmarking

Benchmarking means that the performance of a company is compared with a reference of some kind. There are two alternatives:

1) Comparison with a theoretical optimal utility ("model utility")
2) Comparison with other utilities, also called "best practice"

The major problem with benchmarking of grid companies is the heterogeneity of circumstances under which they operate. Topographic, climatic and other conditions are different for different utilities. It is therefore difficult to construct a model utility or find a best practice utility that is representative.

Data Envelopment Analysis (DEA) is a type of analysis frequently used for benchmarking. It is described later how it is used for efficiency comparisons for Norwegian grid companies. In Sweden there has been an attempt to construct a model utility.

Benchmarking is attractive because the regulation can be based on other input than the utilities' own cost. This is beneficial with respect to incentives.

Quality indicators

In order to avoid cost reduction at the expense of quality, it is relevant to include quality indicators in the regulation procedure. This can either be done by defining absolute quality limits which are not to be exceeded, or by using a penalty depending on the quality level. One example is the Norwegian Compensation for Energy Not Supplied (CENS), which will be described later. This type of indicator gives the utility an incentive to maintain or improve the quality of supply.

9.4 The English price cap regulation

The Office of Electricity Regulation (OFFER) was responsible for the price cap regulation originally introduced in England after the restructuring. The price cap was based on the simple formula: P=PRI - X, where PRI was a general price index and X was an assumed or required productivity development. The formula was in principle the same for the national transmission company (NGO) as for the 12 regional distribution companies (RECs). The X-factor was decided by OFFER. During the first years after restructuring, the development of this factor for the distribution utilities was [5]:

(91-95):   *0 to -2.5*
(95):   *11 to 17*
(96):   *10 to 13*

It is evident that the factor was "generous" during the first years, leading to high profit for the grid owners.[5] After the initial period, when it was in fact assumed negative productivity development, the productivity factor was high. This was

---

[5] Not only was the X-factor beneficial for the grid owners, a considerable cost reduction could also be achieved. This resulted in large profits during the first years after restructuring. It might have something to do with the Government's wish to make electricity distribution interesting for private shareholders. Deregulation in the UK involved privatization.

perhaps due to developments in the initial period. In this case it can be interpreted as a regulatory lag of 5 years.

The responsibility for electricity regulation was later taken over by the Office of Gas and Electricity Markets (OFGEM). The process of deciding the price cap is complex and involves an extensive review of industry performance.

9.5 The Norwegian revenue cap regulation

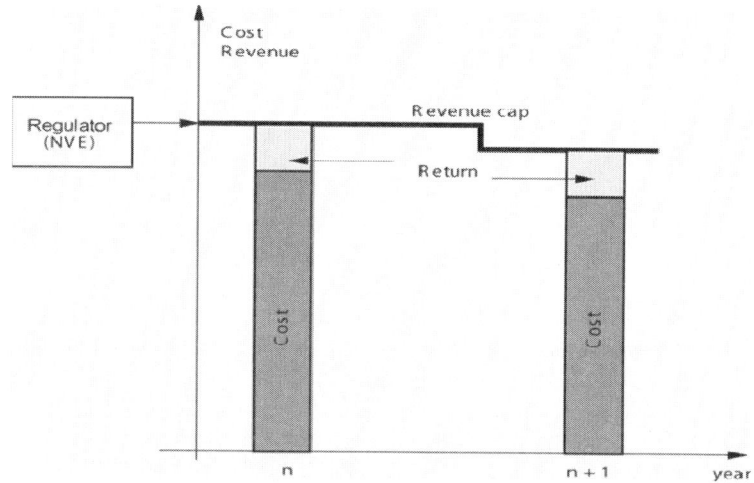

Figure 9.10 Revenue cap regulation

The principle for revenue cap regulation is shown in Figure 9.10. The regulator, which in the Norwegian case is NVE, sets the maximum revenue for each distribution utility. The difference between revenue and cost is the company's profit. The distribution company makes decisions on operation and investments which are decisive for next year's cost. There should in principle no link between these decisions and the revenue cap. Any cost savings should result in higher profit.

Figure 9.10 represents an ideal regulation with respect to incentives. All cost savings give increased profit. In practice, this is difficult to achieve.

9.5.1 Overview

The Norwegian revenue regulation has elements from different regulation mechanisms. There is an element of cost-plus-regulation with a regulatory lag, there is benchmarking and there is a quality indicator.

Figure 9.11 gives an overview of the main factors affecting the revenue cap.

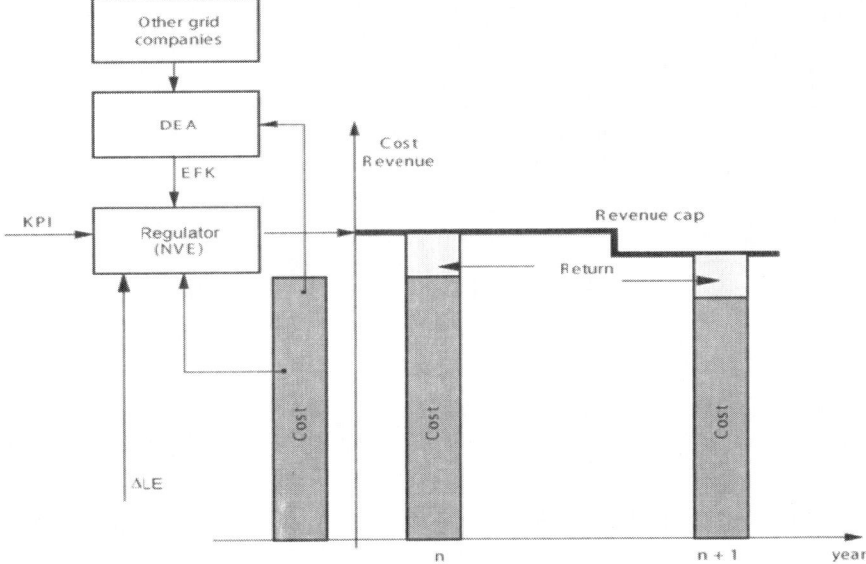

Figure 9.11 The Norwegian revenue cap regulation. Overview.

The development of the revenue cap from year n to n+1 for a regulation period (5 years) is based on the formula:

$$IT_{n+1} = IT_n \cdot \left(\frac{KPI_{n+1}}{KPI_n}\right) \cdot (1 - EFK_{n+1}) \cdot \left(1 + \frac{\Delta LE_{n+1}}{2}\right) \quad (9.1)$$

The different variables are:

1) IT_n: Revenue cap the previous year starting with revenue at the beginning of regulation period.
2) KPI: consumer price index.

3) EFK: Drop in revenue cap due to assumed efficiency gain (based on benchmarking)

4) ΔLE: Increase in volume.

In addition to these four factors included in the formula, the quality in terms of security of supply has an impact on the revenue cap.

We go through the factors one by one:

Revenue at the starting point (IT_1)

Revenue at the starting point is based on a Cost-of-Service mechanism. The first period in the Norwegian revenue cap regulation serves as an example.

The period 1997 – 2001 was the first period with revenue cap regulation in Norway. As the Figure 9.11 indicates, the revenue cap for 1997 was based on costs from previous years including standard return on capital. But requirement for a certain productivity improvement (2 %) compared to previous years was imposed. The same requirement was applied to all utilities this first year.

A similar procedure was undertaken to fix the starting point for the next period (2002 – 2006). As mentioned earlier, Cost-of-Service regulation has basically no incentive for cost reduction. But in this case there is a considerable time lag. Cost savings during the first year could give the utility increased profit up to 5 years ahead.

If the utilities assume that costs during the last years of a regulation period, set the starting point for the revenue cap in the next period, the incentives for cost savings will be reduced. If the assumption is correct, cost savings lead to reduced revenue cap. But the incentive depends on what the utilities believe will happen, not what actually occurs.[6]

9.5.2 Price indexing

As compensation for inflation, the consumer price index is used,

$$KPI_{n+1}/KPI_n$$

---

[6] During the first regulation period in Norway, the regulator, NVE, indicated that that the revenue cap in the second period would be decided in such a way that the utilities would benefit from cost savings in the first. In fact that was not what actually happened, but for the incentives it is important what the companies believe will happen, not what actually happens.

in the income formula.

9.5.3 Efficiency requirement

The factor $(1 - EFK_{n+1})$ represents the assumed annual efficiency improvement in the revenue cap formula.

This efficiency requirement consists of two components: 1) A general requirement applying to all utilities, reflecting potential efficiency improvements for all and 2) an individual requirement based on benchmarking.

The general efficiency requirement was 1.5 % per year during the first period, 1997 – 2001 (except in 1997 when the total efficiency requirement was 2 % for all) and same for the second period, 2002 – 2006.

In addition to this general requirement, additional requirements were imposed on inefficient utilities (as measured by the benchmarking (DEA)). It was required that half the inefficiency was recovered during the period. If the benchmarking indicated 80 % efficiency, 10 % improvement was required during the period, i.e. 2 % per year.

9.5.4 Efficiency measurement by DEA

A benchmarking procedure for Norwegian distribution companies was introduced in order to reveal inefficiency[7]. The efficiency measurement is done by Data Envelopment Analysis (DEA). This is a general type of analysis that is used in different economic sectors. For a detailed description of DEA refer to [4] where there is a general description of the method as well as a detailed description of the model applied to Norwegian distribution utilities.

The DEA model is based on a comparison where the efficiency of every grid company is measured against the best comparable company. In general there is a set of input factors and a set of products for each company. In order to illustrate some features of the model, a two-dimensional (one input, one output) example is used. The example is shown in Figure 9.12

---

[7] What we are talking about here is in fact x-inefficiency, but for simplicity we call it only inefficiency.

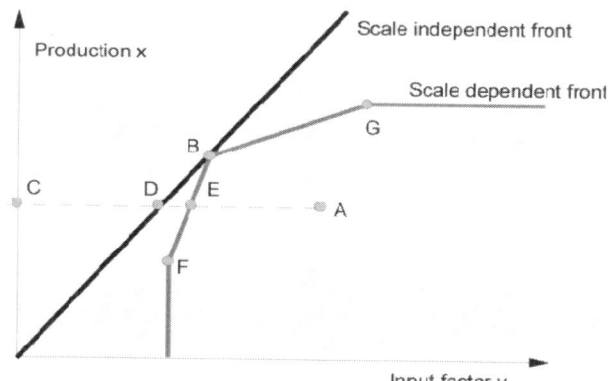
Figure 9.12. Illustration of efficiency measurement with DEA.

For this two-dimensional case, the performance in terms of input and output for utilities can be plotted as shown in the figure. Points A, B, F and G represent different utilities. The efficiency in this two-dimensional case is simply the ratio (x/y). For this simple case it is easy to see that utility B is the most efficient one. Production pr. unit of input (x/y) is higher than for the others. A straight line through the origin and point B represents this maximum observed efficiency. We call this the *scale independent front*. It represents maximum efficiency assuming there is no scale economy. (Scale economy in this case is tied to the size of the utility, see Figure 9.3.). If we assume that efficiency depends on the size of the company, we draw a *scale dependent front* by connecting all the upper-left points in the graph. This convex envelopment line represents the maximum attainable efficiency if scale is taken into account. In both cases we assume that the best observed utility is the best attainable, i.e. 100 % efficiency. DEA is therefore a *best practice* method. We assume that best observed practice is the best attainable.

It is now possible to measure the efficiency of different utilities by comparing each with this best practice front. Utility A can be used as example. From Figure 9.12 we can see that with constant output, it is possible to reduce input corresponding to point D or E if we increase the efficiency of utility A till we reach the scale independent or the scale dependent front respectively. The minimum use of input resources is given by the line CD (or CE) while the actual use is given by CA. So we define the efficiency of utility A as CD/CA or CE/CA depending on whether we use the scale independent or scale dependent front.

In practice there are several input factors and several products. Instead of the two-dimensional problem that can be depicted by a simple graph, there is a multi-dimensional problem. And the fronts are not lines but multi-dimensional surfaces. But Figure 9.12 can still serve as an illustration of the basic principle.

In the Norwegian case, we use the following input and output variables:

- Input:
  - Labour (man-years)
  - Grid losses
  - Capital
  - Other input

- Output:
  - Energy distributed (kWh)
  - Number of customers served
  - Extension of the grid (kilometres of lines etc.)

One of the problems with this model (and one problem with the benchmarking of grid companies in general) is the choice of an output variable for the extension of the area served by the utility. In the Norwegian case, the regulator chose to use the extension of the grid. The problem is that that is not an exogenous variable. It is closely connected to the use of capital, which mainly means grid investments. So grid investments, whether they are needed or not, lead to an increase in that particular output variable.

9.5.5 Revenue to cover costs for increased delivery

Increased delivery, i.e. more kilowatt-hours, leads to increased cost. But the cost increase is not proportional. In the discussion on the scale economy, Section 9.2, we saw that a large increase in delivery could mean to only a small additional cost. This additional cost is covered by the last element in the formula:

$$\left(1 + \frac{\Delta LE_{n+1}}{2}\right)$$

ΔLE_{n+1} is per unit increase in delivery compared with period n. The formula implies that 10% more energy distributed, gives a 5 % increase in the revenue cap. We are using a scale factor of 0.5.

From 2002 new rules were introduced by the regulator. Instead of energy delivered, the number of new customers is now used as an indicator of increased output. In addition, load growth on a national basis is used as an indicator of general growth in consumption.

9.5.6 Compensation for Energy Not Supplied (CENS)

Revenue cap regulation may lead to postponed investments, reduced maintenance and reduced staff. Such actions might in the long run lead to reduced quality of supply, especially to more frequent supply interruptions. As we can see from Figure 9.1, losses for the consumers due to interruptions can be substantial. In order to avoid deterioration of the quality *Compensation for Energy Not Supplied (CENS)* has been introduced in the Norwegian grid regulation. The revenue cap is lowered if customers are disconnected. With the introduction of this compensation arrangement, the economic consequences of this quality aspect can be taken into account in grid planning. This means that for the network owners it will be possible to use technical/economical analysis to find an optimal level of investments and maintenance, minimizing the total costs in which lost load is also included.

NVE introduced CENS from January 2001. The arrangement is based on energy not supplied (for interruptions of duration > 3 minutes) for medium/high voltage end-users. At first, the arrangement differentiate only between residential/agriculture and commercial end-users with different charges. The arrangement also differentiates between incidental interruptions and interruptions with advance warning. The charges are set to 4 NOK/kWh for residential/agriculture and for commercial end-users it will be 50 NOK/kWh for disturbances and 35 NOK/kWh interruptions with advance warning.

The expected average costs caused by non-delivered energy are calculated for each company based on historical data and statistical analysis. The revenue caps for the companies are then initially raised in order to take this average cost for lost load into account. If customer interruption occurs the responsible network company are penalized by a reduction in the revenue cap. If the interruption costs in a specific year are below the company's expected average, the company will get an increase in the income cap (compared with a situation without CENS). If the actual costs due to energy not supplied are above the expected average the income cap will decline. The local network company will provide documentation of affected customers and their estimated ENS, while the owner of the network that caused the interruption is financially responsible for the total amount of CENS associated with the interruption.

There are various challenges of technical, financial, legal and administrative character related to the compensation arrangement: how to record interruptions and estimate energy not supplied, which price to put per unit of energy not supplied, how to divide the responsibility between different network owners etc.

9.6 References

[1] B. Uthus, I. Wangensteen: "Monopolkontroll av nettvirksomheten - en drøfting av regime og incentivstrukturen" EFI TR A4944 februar 1999

[2] I.Wangensteen, E.Dahl: "An Investigation of Distribution cost by means of Regression Analysis" IEEE/PES , 1989

[3] E. Hope: "Studier I markedsbasert kraftomsetning og regulering", FAGBOKFORLAGET Bergen 2000

[4] Sverre A. C. Kittelsen: "Effektivitet og regulering i norsk elektrisitetsdistribusjon". SNF- rapport nr. 3/1994

[5] Masayuki Yajima: "Deregulation Reforms of the Electricity Supply Industry" Quourum Books, 1997

10 Environment policy

10.1 Introduction

The power system has great impact on the environment. Electricity generation from fossil fuel causes emissions as well as depletion of the non renewable resources. Hydropower causes damage to lakes and river courses. Nuclear energy has its own environmental problems. The transmission and distribution of electricity is also problematic. Overhead lines are detrimental for the landscape and health impacts of overhead lines are being discussed. Electricity consumption on the other hand has little or no environmental impact.

We will not go into details about the different environmental consequences here, rather concentrate on different policy measures used to influence the system in an environmentally friendly direction. We will mainly concentrate on the generating side.

What the most environmental friendly technologies are, can of course be discussed. It is to some extent based on individual judgment. Some people find a wind turbine attractive and some find it damaging to the landscape. But normally there are two criteria for what is categorized as "green electricity": 1) It should be based on a renewable source and 2) it should have low or zero emissions (in particular for CO_2). Based on these criteria, the following sources are normally included:

- Wind power
- Bio energy
- Small hydropower
- Solar energy
- Waves and tidal power
- Geothermal power

Electricity from biomass of course creates emissions, but we are assuming the CO_2 are recirculated through the assimilation of CO_2 by forests, so it gives no net addition. Other emissions from biomass can have serious local effects, but despite this, biomass is normally regarded as "green".

Peat is a fuel somewhere between fossil fuel and biomass. In some cases it has been included as a "green" source of energy because the thickness of a peat layer is slowly increasing. Today the common view seems to be that peat cannot be included because the recirculation is very slow and because there are serious local impacts from extensive exploitation of peat.

With respect to small hydropower, definition is always a problem. We have to set an upper limit to what is regarded as small. According to Norwegian definition (from NVE) the limit is 10 MW[1].

The categories listed above are what we call *new renewable sources* or *green electricity*.

10.2 Policy instruments, overview

The choice of tools for the promotion of environmentally good solutions depends on the regulatory environment. If the supply system is based on central planning, direct control (Command-and-control, CAC) is the most appropriate instrument. It is possible on the basis of political guidelines or preferences to use multi-criteria planning in order to come out with an optimal plan for system expansion where environmental aspects are taken into account. Final decisions are taken by central decision-makers, in many cases political authorities, and implemented on a CAC basis[2].

In a deregulated environment, which is the type of environment we will assume here, planning of investments and operation of the power supply system will be done by each individual company on an economic basis, which means maximum profit. Under these circumstances, different instruments can be used.

We start with a short overview. More detailed descriptions of the most important and currently most interesting instruments follow later.

10.2.1 General instruments

These instruments are directed towards the economy in general or towards a broader part of the economy than just the power supply sector.

---

[1] NVE distinguishes between three classes: micro < 100 kW, mini < 1 MW and small < 10 MW
[2] Multi-criteria planning is a complex subject. Depending in how the preferences are expressed, a multi-objective or an ordinary economic optimization model can be used.

Restrictions (CAC)

Restrictions of more or less general nature are frequently used for environmental purposes. Typical examples are restrictions on certain geographical areas. There are protected areas where several types of exploitation are prohibited. National parks are typical examples.

Other examples of restrictions are limits (caps) on polluting emissions to air, water or land.

Market-based instruments (MBI)

Typical examples of market-based instruments are taxes or subsidies (corresponding to negative and positive externalities respectively) imposed on producers.

Combined instruments

Typical examples are so-called *cap-and-trade* instruments where a cap is imposed on a certain type of emission, but it is opened up for trade of emission permits. Tradable emission quotas are important for the achievement of the greenhouse gas targets determined by the Kyoto Protocol. Emission quotas are fixed for each country and it is up to the authorities to issue emission permits to electricity producers and other firms. If the sum of the emission permits is less than the total national quota, the rest can be sold either to national firms or on the international market. Firms with more emission permits than needed, can sell the surplus on the market. Firms with less permits than their actual emission, will have to buy. We return to this in more detail later.

10.2.2 Sector specific instruments

This means instruments confined to this particular sector, i.e. the power supply sector. The instruments can be classified in basically the same way as the general instruments.

Restrictions (CAC)

This is the traditional way to take environmental aspects into account in the power sector. In a hydropower system (typically in Norway) restrictions are imposed on a national level and on the individual power plants. On the national level we have different plans imposing restrictions on hydropower development. Certain rivers and lakes are protected. Restrictions are also imposed on operation of individual plants. There are restrictions on upper and lower levels on reservoirs and in many cases on minimum and maximum flow in a river course.

Thermal power plants have restrictions on emissions, typically for SO_2, NO_x.

Market-based instruments (MBI), (taxes, subsidies, in-feed-tariffs)

Market-based instruments are extensively used in the power system.

In most countries we find a general tax on electricity consumption. In Norway it has been about 10 øre/kWh during the last few years. In Denmark the tax level is close to one DKK/ kWh. As will be shown later, a consumption tax is not the best solution from an environmental point of view. But there are also fiscal motives for such taxes. Subsidies are in many cases used in order to promote green electricity. They can be used on a case by case basis, or it can be used on a long-term more predictable way.

Use of feed-inn tariffs is a special form of subsidy. In contrast to usual subsidies, this support is tied to the grid tariff. It can be regarded as a negative tariff for use of the distribution and transmission grid. The grid companies are simply instructed by the government (or more generally by the political authorities) to grant a certain economic support to power input from green producers. The extra cost to the grid company has to be covered by the general grid tariff.

The use of feed-in tariffs has been a dominant support mechanism for renewable electricity producers in the last decade. Feed-in tariffs function as a form of fixed price that the producer is guaranteed to receive for every unit of green electricity that it feeds into the grid. The tariffs usually vary depending on the type of technology and time of production. This instrument gives high security for investments, but does not enhance price efficiency. Germany's feed-in tariffs for green power is a good example. It has resulted in large investments in renewable generation capacity.

Tendering systems

If a government has set up a target for a certain amount of green electricity, it is possible to use a tendering system to acquire that amount at a minimum cost. Potential generators of green electricity are invited to compete either for a certain financial budget or a certain capacity of renewable electricity generation. The public body acts as a joint buyer on behalf of the consumers. Different technologies can have separate tenders, and the cheapest bids per kWh are awarded the contracts within each technology.

The green electricity is sold to the local utilities, which pay the bid price per kWh. A premium equalling the difference between this price and the market price is reimbursed by a fund, which is financed by a non-discriminatory levy paid by all electricity consumers.

As the term implies, competitive bidding enhances strong competition between investors. However it may prove discriminatory against small investors.

Certificates

Several countries are either in the planning process or have already implemented a system for supporting renewable electricity production through trade with green certificates. This system is explained in detail in Section 10.6. Competitive bidding and green certificate systems have much in common, but one of the main differences is that with a certificate system consumers will not be represented jointly by a public body, but are instead allowed to act individually on the market.

10.3   Cap or tax

The damage caused by pollution increases with increasing quantity of emissions. The cost of abatement, either by curtailing the production causing emissions or by purification measures, will in general increase with decreasing emissions. The trade-off between the two: environmental costs of emission (to society) and cost of reducing emissions (normally a cost to firms causing the emission) is shown in Figure 10.1. The sum of the two cost components is also shown in the figure. The optimal solution is the point where this total cost has its minimum. This is the point where the marginal abatement cost equals the marginal environmental cost as indicated in the lower part of figure.

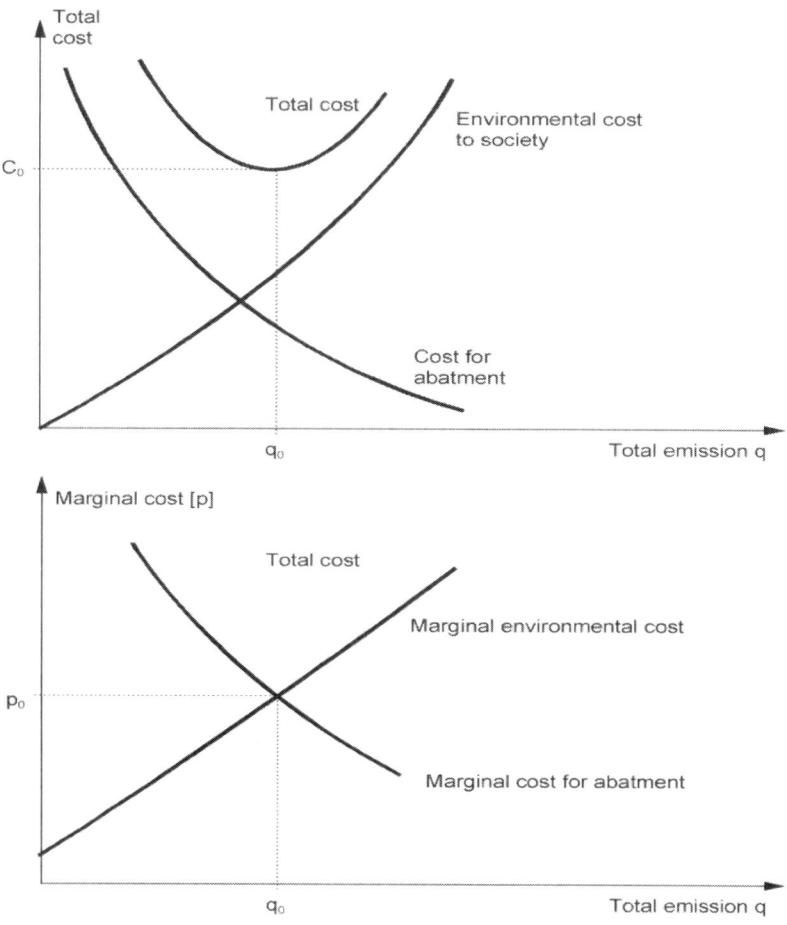

Figure 10.1. Cost of emissions (to society) and cost for abatement of emissions. Total cost is minimized in the point where marginal environmental cost equal marginal abatement cost.

From the figure we can see that in order to obtain the optimum solution, there are two alternative instruments: either we can choose an environmental tax or we can choose a cap on total emissions. If we impose an environmental tax, p_0, the companies will reduce emissions rather than pay this tax as long as the marginal abatement cost is lower than p_0. When the abatement cost gets higher, they will pay the tax and the emission will not be reduced below the level q_0. If we impose a cap

q_0, the companies will have to curtail emissions and that will entail abatement cost on the companies until we reach level p_0.[3]

In short: We can impose a tax leading to an optimal quantity or we can impose a cap leading to an optimal marginal cost. The choice is to some extent a practical question. But it can also be important how well the abatement or environmental costs are known and how variable the costs are with quantity. Suppose for instance that the marginal abatement cost is almost constant over a wide quantity range and only approximately known. Environmental tax will in that case be a poor instrument. There is a risk that the tax is too low and have no impact. Or the tax can be too high and emissions reduced almost to zero, which is also a non optimal solution. If the marginal abatement cost falls steeply with decreasing quantity, an emission cap can be an unsuitable instrument. If we miss the target, there is for instance a risk that high abatement cost must be used in order to reach the cap.

10.4 Environmental taxes and subsidies

10.4.1 General

As mentioned in Chapter 2, the use of environmental taxes was an idea introduced by the English economist Pigou as early as the 1920s. It is of course a problem to assess the consequences of pollution or other externalities correctly in economic terms. However, if we assume we have a correct assessment, we can put a tax on the producers as shown in the Figure 10.2.

---

[3] We assume that abatement measures are implemented in merit order. That can be difficult to achieve if there are many companies involved. One way to solve that problem is emission permits trading. We return to that in Section 10. 5.

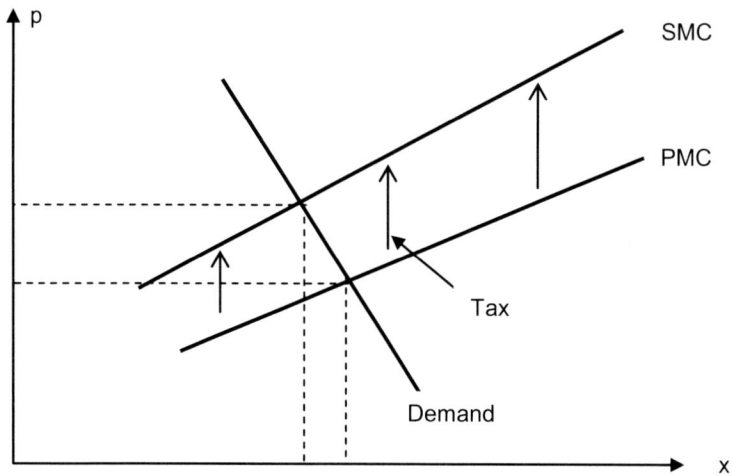

Figure 10.2 Taxation reflecting external cost.

The tax will increase the marginal cost of the producers, from PMC (private marginal cost) to SMC (social marginal cost), and thereby cause a shift in the supply curve as indicated in the figure. The price will increase and the quantity decrease accordingly.

If the external cost is equal for all producers (per unit produced), the tax can either be put on the producers or the consumers. It is easy to see that the effect will be the same. See figure 2.10 for an illustration of that. However, some producers cause more pollution than others. Consequently, the general rule is that environmental taxes should be put on the producers. The so-called "Polluter Pays Principle" is a well-known principle in this context. The firm causing the pollution should be taxed according to the damage.

These Pigouvian taxes have two effects. Firstly, the taxes cause an efficient allocation of resources as indicated in Figure 10.2. The price after taxation equals the social marginal cost. Secondly, the taxes give public income. In general taxation is a problem because it is normally causing economic inefficiency. However, if a Pegouvian tax is introduced, other taxes can be reduced accordingly and the corresponding inefficiency is also reduced. Due to this, it is claimed that environmental taxation gives a double dividend.

10.4.2 Operational consequences

In order to give an illustration of the short-term effect of a Pigouvian tax on a power generation system, we look at a merit order supply curve. See Figure 10.3.

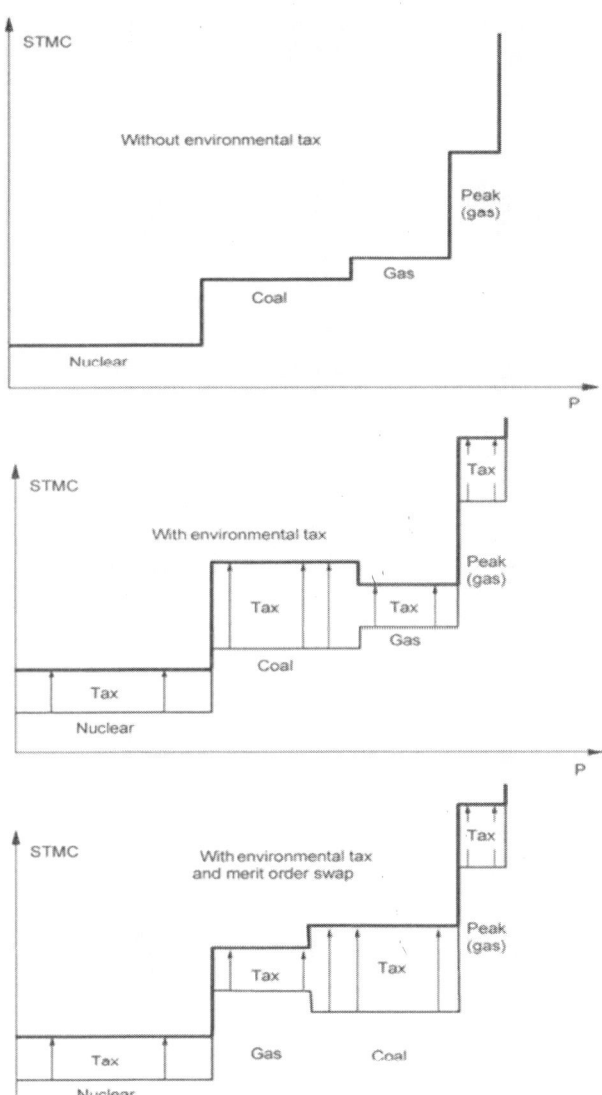

Figure 10.3. Supply curve for a thermal system without and with environmental tax

Chapter 10: Environment policy

The example chosen is a typical thermal system with nuclear as base load, coal and gas as low/medium load and gas as peaking capacity. The upper figure shows the merit order supply curve without taxes. We then impose a tax which is different for the various technologies. That is shown on the next curve. We have assumed tax levels reflecting the externalities of the different technologies, implying that the tax on coal fuelled generation is about twice that of generation based on gas. We see that with the costs and tax levels used in this example, the cost for generation based on coal will be higher than that based on gas. The consequence is that there will be a swap in the merit order curve. The final merit order curve is shown at the bottom of Figure 10.3.

We notice that there will be a reduction of the external cost (emissions) for two reasons:

1) There will be reduced demand due to higher prices and 2) there will be a shift from more polluting to less polluting technology i.e. from coal to gas. There will be a reduction in the use of coal and the use of gas will probably increase. Nuclear as a base load technology will hardly be affected.[4]

10.4.3 Long term impacts

We will here look on long-term impacts of environmental taxes or subsidies. Our point of departure is in this case the long-term supply curve (reflecting the Long-Term Marginal Cost, LTMC) of what we call conventional power supply technologies. In addition we have a similar supply curve for green technology, see Figure 10.4. Supply curves for conventional technology (with and without environmental tax) and supply curve for green electricity. With the costs indicated in this figure, green electricity is not competitive.

---

[4] In order to be able to estimate the consequences, we need information about the demand side. We need to know the price elasticity of demand and we need information about the level and variability of demand, for instance in the form of a load duration curve.

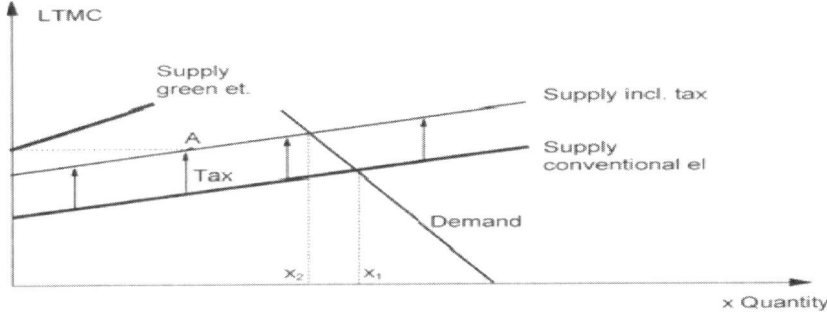

Figure 10.4. Supply curves for conventional technology (with and without environmental tax) and supply curve for green electricity.

If we introduce a tax on conventional technology supposedly reflecting external cost for that technology, we get a vertical shift of the supply curve. That new curve represents the social marginal cost for that technology. But we see that with this new curve, the green technology becomes competitive. From point A on the curve, green technology can compete with conventional technology. That leads to a new aggregated supply curve, see Figure 10.5. Below point A we have only conventional technology. Between points A and B we have a mix of conventional and green (sometimes called grey) technology. The balance point (x_0, p_0) represents the social optimum (provided the taxes are reflecting true social costs and, as we have implicitly assumed, and there are no external costs tied to green electricity) Figure 10.5.

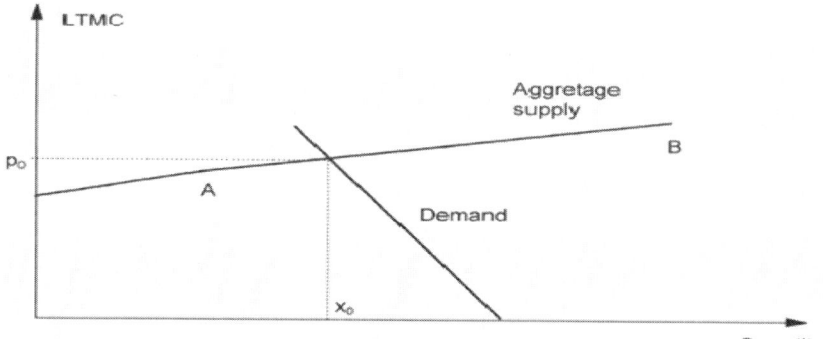

Figure 10.5. Aggregated supply including green electricity.

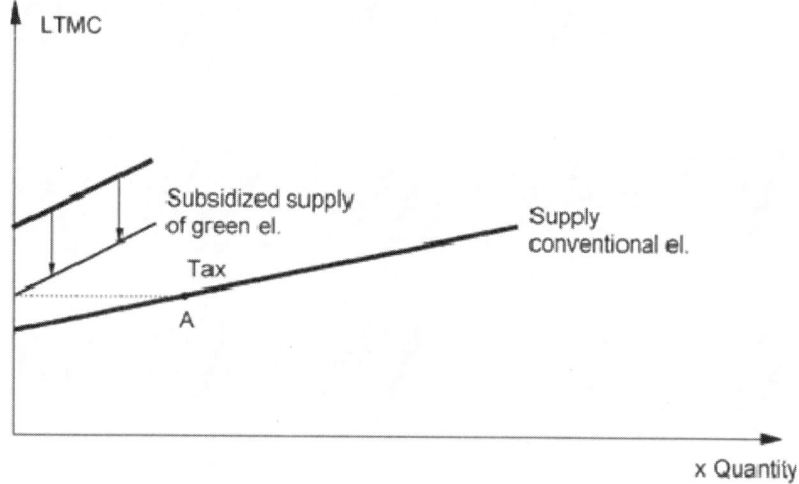

Figure 10.6. Impact of subsidies on green electricity.

We next look on the impact of subsidies. Our point of departure is the same, see Figure 10.6. Green electricity is not competitive at the outset, but after introduction of subsidies, the marginal cost will decline and green electricity can compete with conventional. The competitive situation is in fact the same as in the case with environmental tax (we have assumed that the subsidy and tax level are equal). The difference is that competition is now going on at a lower price level. The two solutions are compared in Figure 10.7.

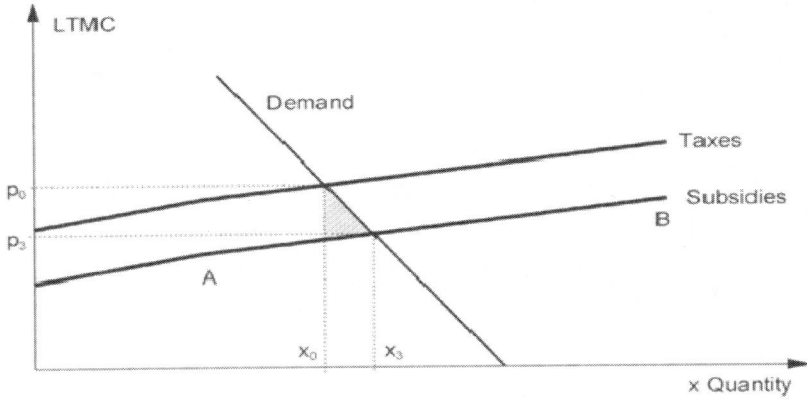

Figure 10.7. Demand supply balance for two cases: 1) taxes on conventional technology and 2) subsidies on green technology.

We see that subsidies lead to lower price and there will be a loss of economic efficiency due to the fact that the consumers will not see the full social cost. The economic loss (welfare loss) of the subsidy alternative compared with the tax alternative is indicated by the shaded area in the figure.

10.5 Cap and trade

To introduce a cap on total emissions can be a useful way to limit environmentally damaging emissions. Section 10.3 discussed the use of environmental taxes versus introduction of a cap on emissions. Here it is shown that an emission cap should be followed up with trading arrangements in order to achieve the limitation of emissions in the most cost-effective way possible.

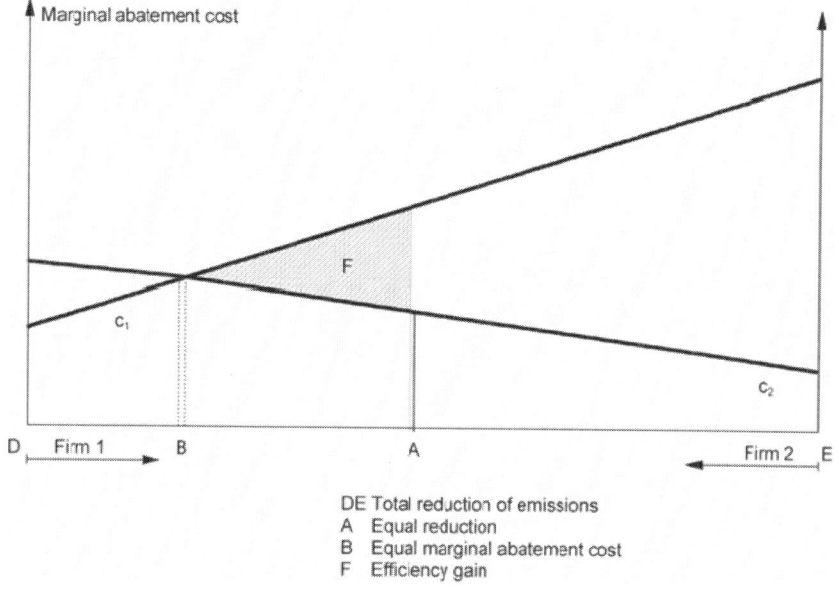

DE Total reduction of emissions
A Equal reduction
B Equal marginal abatement cost
F Efficiency gain

Figure 10.8. Benefit of emission trading

If an emission cap or a quota is introduced, that quota has to be divided between the different emitters. Emitters can in this case mean countries or individual firms. There are different procedures for this[5], but in general one will decide on individual emission permits without knowing the abatement costs for the affected entities.

The advantage of trade is illustrated in Figure 10.8 above. We suppose that the total emission is to be reduced by the quantity DE and this reduction is divided between two firms, Firm 1 and firm 2 as shown in the figure. We suppose the reduction obligation is shared equally between the two, point A in the figure. Reduction by Firm 1 is represented by moving from left to right and the corresponding marginal abatement cost is c_1. Reduction by Firm 2 is represented by moving from right to left and the corresponding marginal abatement cost is c_2.

We see that at point A, the abatement cost for Firm 1 is higher than that for Firm 2. It is therefore beneficial for Firm 1 to buy emission permits from Firm 2. It is

---

[5] One well-known example is the Kyoto agreement for distribution of CO_2 quotas among industrial nations. The national quotas were settled through a long and complex negotiation process. These quotas are next distributed among industrial firms within each nation. Again there is complex process where strong economic interests are at stake. In many cases emission permits are allocated (free of charge) to firms based on their historical emissions record. This practice is often called "grandfathering".

profitable for Firm 2 to reduce emissions and sell to Firm 1 until point B is reached. That is economically profitable for the two firms and for society. The efficiency gain is represented by the area F.

In this example the emission reduction is initially shared equally between the two firms. This is not the only solution. There are different ways to divide a reduction or a quota between the affected parties. This is subject to negotiations and it can be difficult because commercial interests can be heavily involved.

10.6 Tradable Green Certificates

In liberalized markets cost-efficiency, low consumer prices and environmental policy goals must be achieved. One of the obvious difficulties in introducing renewable energy technologies, also referred to as green technologies, is the fact that many of these are still immature and not as yet competitive in a liberalized market. If the share of renewable energy supplies is to be increased, it will be necessary to create a framework within which investing in green technologies is considered a desirable option, or maybe rather *the* desirable option. This must be done without laying unbearable cost burdens on society.

This is the main idea behind creating a market for tradable green certificates (TGC), which is a market-based support scheme for renewable energy technology. The TGC trade can comprise electricity, heat and other use of renewable energy sources. However, for reasons that will be explained later in this chapter, the main focus in the following will be a TGC market for electricity.

A TGC market is a purely financial market operating in parallel with the physical electricity market. A green power producer sells electricity into the grid, and at the same time receives a number of green certificates corresponding to the amount of green electricity sold. The certificates are tradable financial assets. By selling a green certificate the producer will get coverage for the additional expenses that it has had for producing green rather than grey electricity. Figure 10.9 shows a principle drawing of how the TGC market operates in parallel with the electricity market.

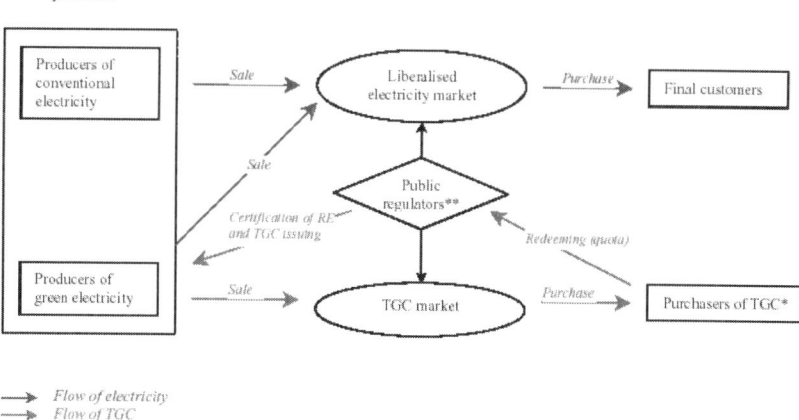

Figure 10.9. The TGC market in parallel with the electricity market. Source: UCL.

The green certificate in itself has two purposes. First it functions as an accounting system that verifies whether a demand has been met or not. Alternatively, if there is no demand, it measures the amount of electricity produced from renewable energy sources. Second, green certificates facilitate trade through the creation of a green certificate market that functions independently of the market for trade in electricity as a commodity.

10.6.1 The goal of TGCs

The goal of green certificate trade can be summarized in six main objectives:

1) Certify, support and make visible the use of renewable energy.

2) Stimulate the development of *new* renewable energy technology in a cost efficient way.

3) Stimulate technical development within certain green technologies in a controlled direction and speed.

4) Increase security of supply, both with regards to domestic energy production, and in renewable rather than non-renewable energy.

5) Transfer some of the support expenditure of renewable energy technologies from the authorities to the market.

6) Stimulate research and development within RETs, as well as the job market within the energy sector.

A main objective of the green certificate system is to increase the share of green electricity on the market. If the green technologies were already economically able to compete with conventional technology without additional financial support, then a green certificate market would be without purpose.

10.6.2 Supply and demand of green certificates

A RE target must be politically determined and controlled. The target can be set as a percentage[6] or absolute-amount obligation in either kWh consumed/ supplied or in MW installed capacity. The target can increase over a predefined period of time, to allow a gradual increase in the expansion of RET.

A demand for green certificates can arise for several reasons. There may be a voluntary demand by consumers, e.g. by green pricing and labelling. The government can also impose an obligation on consumers or other actors in the electricity supply chain (generators, distributors, suppliers) to generate, transmit, deliver or buy a certain amount of green certificates, thereby creating a demand. Alternatively the government itself may choose to act as a buyer of green certificates, for instance to secure a minimum price. In practice demand might come from a combination of these sources. In a voluntary market the incentive to buy certificates may be created if purchases qualify for receiving a benefit (e.g. tax exemption), while in the case of an obligation the incentive is given through imposing a penalty for non-compliance.

Competition between producers of renewable energy will ensure that the renewable energy targets are met in a cost-effective way. The competition will lead to declining costs of renewable electricity generation and an increasing supply of green certificates. The TGC price depends on supply and demand. If the supply is low, the price will be high, giving an incentive for investors to invest in, and producers to produce more, renewable electricity generation. Those producers who are able to provide the cheapest renewable electricity will be able to sell their certificates.

---

[6] Every percentage obligation is of course in the end translated into an absolute amount.

Why not include heat in the certificate market?

Several countries have considered issuing TGCs for renewable heat, but so far very few have done so.

Heat energy has several characteristics that are fundamentally different from those of electric energy. The difference in their way and ability to be transported and measured makes trade with green heat as a commodity differ from trade with green electricity. Once generated, heat is a local resource, which must be used within a limited distance. The fuels may be transported to the consumer prior to the heat generation, however at added costs. This puts a limit on the market's size, consequently limiting the ability of the green certificate system to enhance competition and lead to cost-efficient solutions. Another obstacle may be to find an accurate method of measuring and quantifying the energy value in heat. These are characteristics that are very different from electricity, which is both easy to measure and to transport over long distances with low losses.

10.7 TGCs and standard economic theory

The idea behind a TGC market is that the competition among producers of electricity based on RE ensures that the TGC price, given a certain demand, shall reflect the actual cost difference between producing green and conventional ("black") electricity. This will function as a clear price signal to the authorities and producers as to what the marginal costs are to survive on the green market. It should however be noted that this requires a transparent market. If constraints like for instance price-caps and minimum prices are included in the TGC system design, the transparency is reduced.

Figure 10.10 shows the equilibrium price of TGCs in a static obligation-only market, with market transparency and immediate adaptation of production to supply. MC renewables is the marginal production cost curve of electricity from RE sources. B is the set RE target, with corresponding marginal cost mc^*. If the market price for electricity corresponded to P_E drawn in Figure 10.10, part of the renewable electricity (up to A) would be compatible with conventional electricity without the support from green certificates. However, in order to achieve the RE target (B), an additional amount must be added per unit. Given a market price (including general taxes) for electricity of P_E, the certificates will be sold for $P_C = mc^* - P_E$.

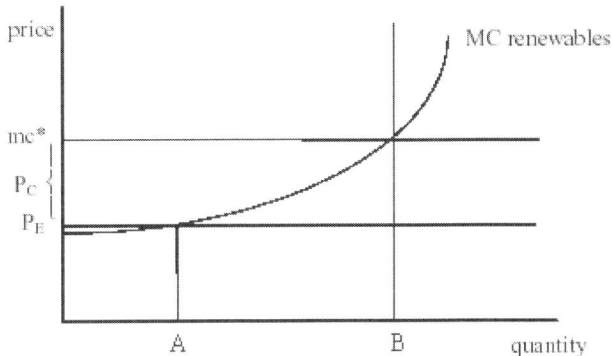

Figure 10.10. Green certificates market. Source: Öko-Institut, SPRU.

In other words, the marginal green technology can be produced at a cost equal to $P_E + P_C$. Only green producers who are able to deliver electricity at a lower cost than this, will be able to profit from delivering green electricity to the grid. The total profit is equal to the area between the mc^* line and the marginal cost curve, MC renewables. The market has thus in its own way rewarded the most cost-efficient technologies and investments.

Assuming that the obligation is imposed on the consumer, a certain share of their electricity consumption must be produced from renewable energy sources. This quota (α) may increase over a time period, in order to reach a final goal in a certain year. The consumer price for electricity (excluding transmission costs) will then equal $P_E + \alpha\, P_C$. The consumer's price elasticity on electricity will thus be divided between the electricity price and the certificate price.

10.7.1 A socio-economic approach

In a market consisting of both conventional and renewable electricity producers, the introduction of a TGC market will inflict a change in the economic surplus, affecting the electricity consumers and the producers of conventional and renewable electricity in different ways. In the following section a step-by-step analysis will show the economic consequences that can be expected from introducing a TGC trading scheme into the electricity market. It should be noted that the graphs do not correspond to a given case, but are simply for illustrative purposes.

Figure 10.11 shows a standard supply curve (S) and demand curve (D), in this case made linear for simplicity. Note that the upper dotted part of the supply curve stands for the renewable electricity supply, and is for simplistic reasons assumed to be more expensive than all thermal electricity production. In a liberalized market the curves intersect at the point of market equilibrium, i.e. the point at which the marginal cost equals the willingness to pay at a common price (p_1) and volume (x_1). The economic surplus is shown as the shaded triangular area between the two curves, of which the area above p_1 is the consumer's surplus and the area below is the producer's surplus.

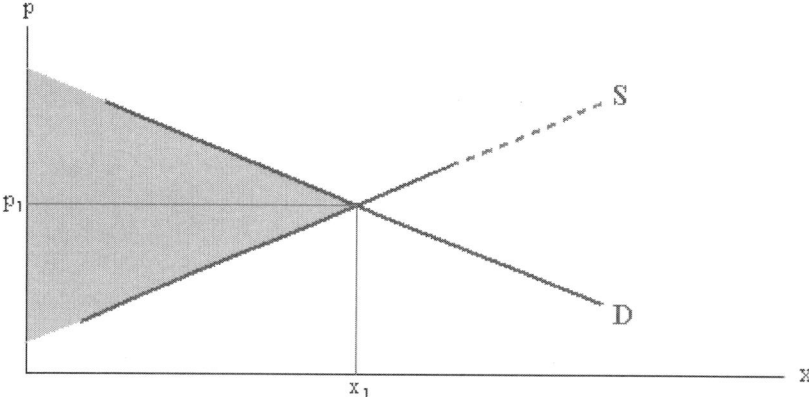

Figure 10.11. The Equilibrium point (x_1, p_1) between the supply (S) and demand (D) curve.

When imposing a mandatory TGC market, a certain share (α) of renewable electricity is integrated into the supply curve, causing a shift to the right (S+) corresponding to $x_n*(1+\alpha)$, as shown in Figure 10.12. This relation causes S+ to be non-parallel to S. Note that the S+ curve does not include the added cost of renewable electricity.

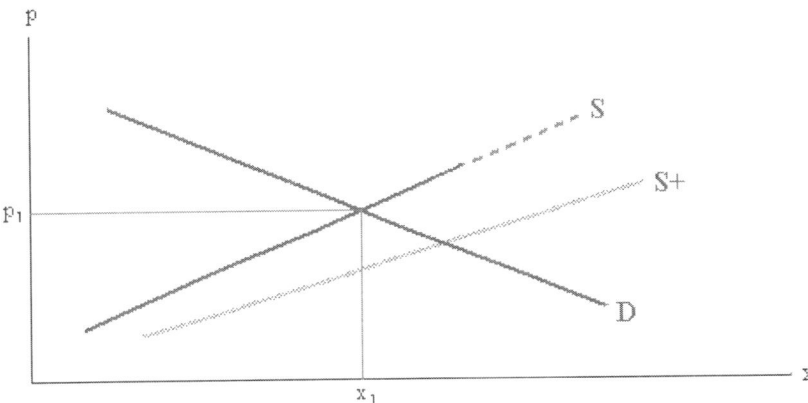

Figure 10.12. Integrating renewable generation into the supply. S+ comprises the new supply curve, but does not include the additional cost that the renewable share will inflict.

In reality the consumer will experience a higher cost because of the mandatory share of TGCs that are included in the supply. This results in a parallel shift upwards, as indicated by S^* in Figure 10.13. S^* intersects with D in (x_2, p_2), thus determining the consumer price (p_2) and total amount of consumed electricity (x_2). Note that in this example the introduction of a mandatory green share into the demand causes a fall in consumer price, and consequently an increase in the total electricity consumption. This counterintuitive effect is pointed out in several analyses.

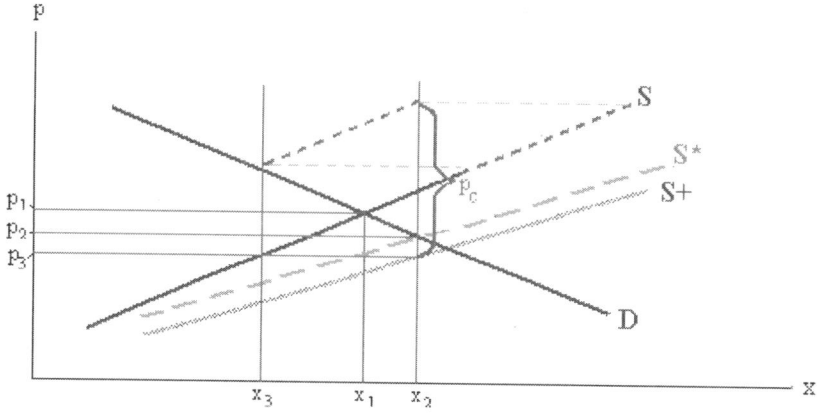

Figure 10.13. Impact of the additional consumer's share of green electricity.

The new sales price in the traditional market falls to p_3, and the total amount of sold conventional electricity is reduced to x_3. The amount of green electricity produced and sold in the market is $(x_2 - x_3)$.[7] The green supply curve can be horizontally transposed and placed above the amount of green electricity for illustrative purposes. The share of green electricity (α) relative to the total electricity consumption is presented in Equation (10.1).

$$\alpha = \frac{(x_2 - x_3)}{x_2} \qquad (10.1)$$

The certificate price, p_c, equals the excess amount that is needed, in addition to the income from the regular sale on the spot market ($p_3^* \, x_2$), to cover the extra expenses due to producing green rather than conventional electricity. The consumer price, p_2, is derived as shown in Equation (10.2).

$$p_2 = (p_3 + \alpha p_c) \qquad (10.2)$$

---

[7] Again it is stressed that the figures are strictly hypothetical, and do not correctly illustrate to the share of renewable electricity relative to conventional electricity.

The fact that the consumer is not faced with the actual marginal cost of renewable electricity production, but rather a fraction of it, interferes with the normal relations between cost and consumption. Hence the increase in consumption from x_1 to x_2.

When looking at what happens to the socio-economical surplus, the first noticeable change is the shift in the consumer's and producer's surplus, as can be seen by comparing the shaded areas in Figure 10.14 with the shaded areas in Figure 10.15 and 10.16.

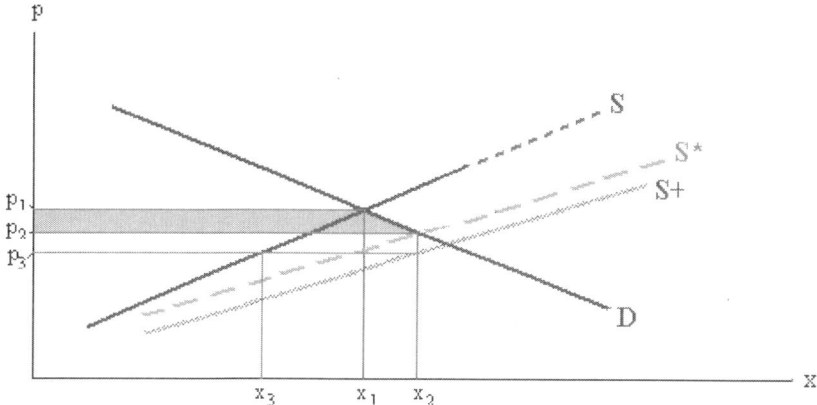

Figure 10.14. The increase in consumer's surplus, indicated by the grey shaded area.

There is an increase in the consumer's surplus due to the reduction in consumer price, shown as the grey shaded area in Figure 10.14. In addition the producers of green electricity now gain a position in the market, giving them a producer's surplus equivalent to the triangular grey shaded area above the green supply curve in Figure 10.15. The producers of conventional electricity, on the other hand, experience both a loss in demand for conventional power and a price fall on the spot market. The result is a loss in producer's surplus, corresponding to the shaded grey area between p_1 and p_3.

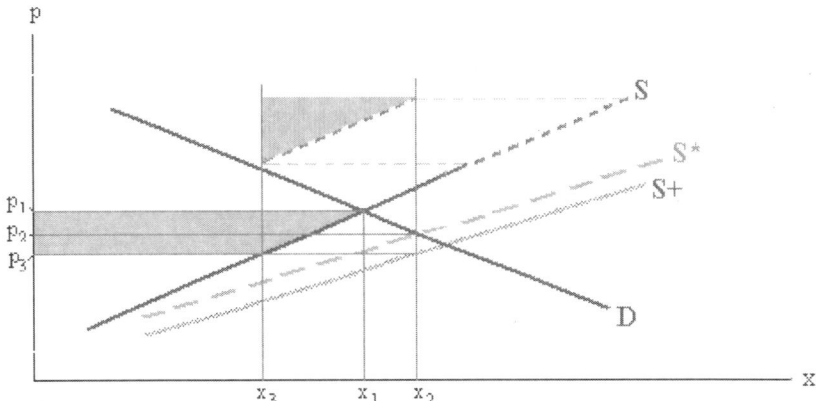

Figure 10.15. The shift in producer's surplus from conventional to thermal production.

The producers of conventional electricity experience a reduction in surplus represented by the lower shaded grey area to the left of the supply curve (S), between p_1 and p_3. The renewable electricity producers on the other hand now receive a surplus corresponding to the upper shaded grey triangular area.

The new socio-economic surplus after including a green share in the electricity market is illustrated in Figure 10.16.

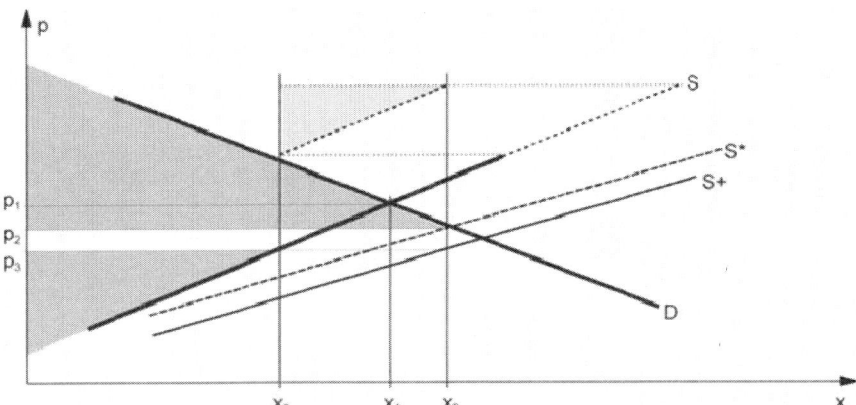

Figure 10.16. The total socio-economic surplus after including a green electricity share on the electricity supply, shown as the grey shaded area.

Another matter of interest is finding the area in the economic model that equals the loss in economic surplus due to the introduction of the TGC market. This loss equals the value society pays for the environmental benefit of renewable electricity.

To pursue this quest, the supply and demand graph is divided into smaller areas, so as to try to keep a system in the following reasoning (see Figure 10.17).

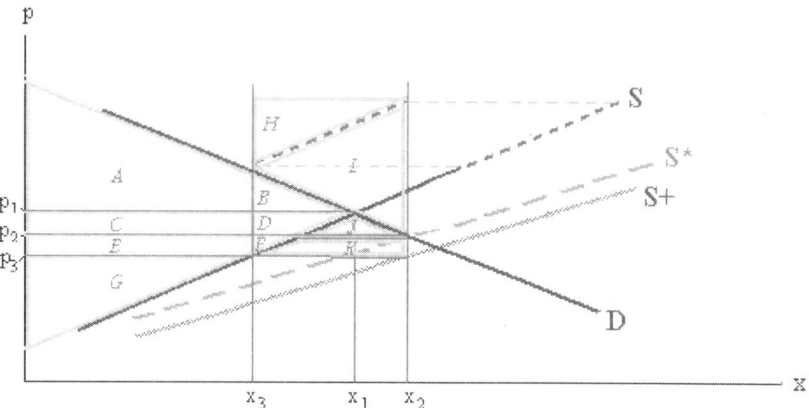

Figure 10.17. In the process of evaluating the total loss in socio-economic surplus, the supply and demand graph is divided into small areas, labelled with the letters A to K.

The socio-economic surplus prior to the introduction of the TGC market corresponds to Equations (10.3) – (10.5):

Consumer's surplus = A + B (10.3)

Producer's surplus = C + D + E + F + G (10.4)

Socio-economic surplus without TGC market = A + B + C + D + E + F + G (10.5)

After the TGC market has been implemented, the evaluation of the changed socio-economic surplus is given as shown in Equations (10.6) – (10.9).

Consumer's surplus = A + B + C + D + J (10.6)

Conventional Producer's surplus = G (10.7)

Green Producer's surplus = H (10.8)

Socio-economic surplus with TGC market = A + B + C + D + J + G + H (10.9)

In addition there is the following relation:

Cost of purchasing renewable certificates = Income from selling renewable certificates

$$E + F + K = H + I + B + D + J + F + K \qquad (10.10)$$

The cost of purchasing renewable certificates is shown in Figure 10.18 and the income from selling renewable certificates is shown in Figure 10.19.

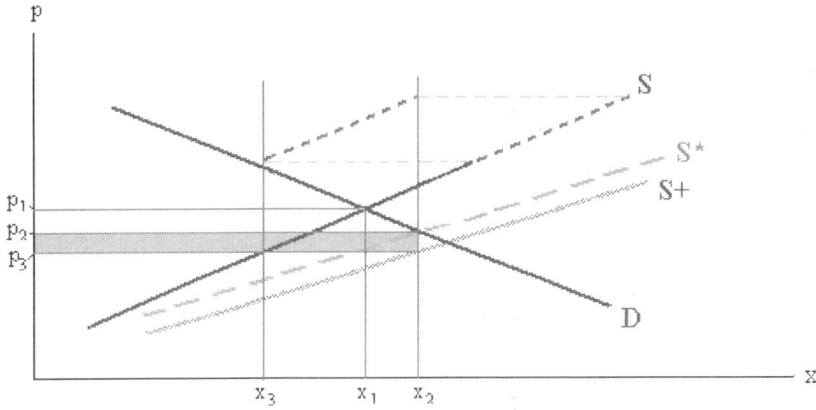

Figure 10.18 The consumer's cost for purchasing renewable certificates, shown as the grey shaded area.

Chapter 10: Environment policy

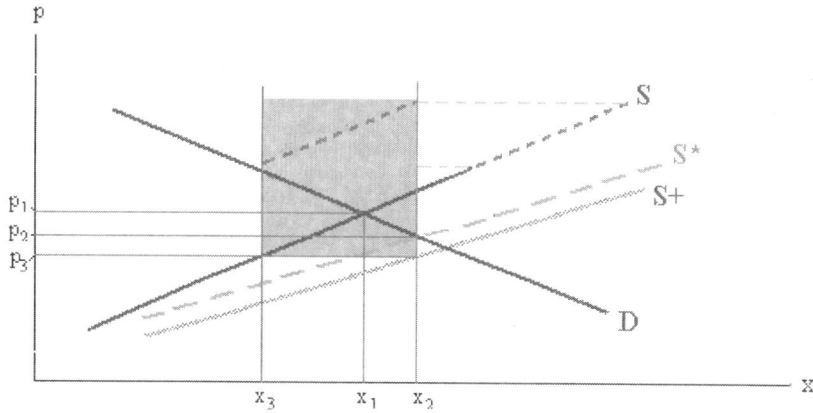

Figure 10.19. The renewable electricity producers' income from selling green certificates, shown as the grey shaded area.

The change in consumer surplus is thus:

Change in socio-economic surplus

= socio-economic surplus with TGC market – socio-economic surplus without TGC market

$= (A + B + C + D + J + G + H) - (A + B + C + D + E + F + G)$

$= J + H - E - F$ (10.11)

Equation (10.10) can be reorganized to the following:

$(J + H) = E - I - B - D$ (10.12)

Combining Equation (10.11) with Equation (10.12), gives the following relation:

Change in socio-economic surplus $= (E - I - B - D) - E - F = -I - B - D - F$ (10.13)

Since all of the components are negative, the change in socio-economic surplus is bound to be negative. The area corresponding to this loss in socio-economic surplus corresponds to the grey shaded area in Figure 10.20.

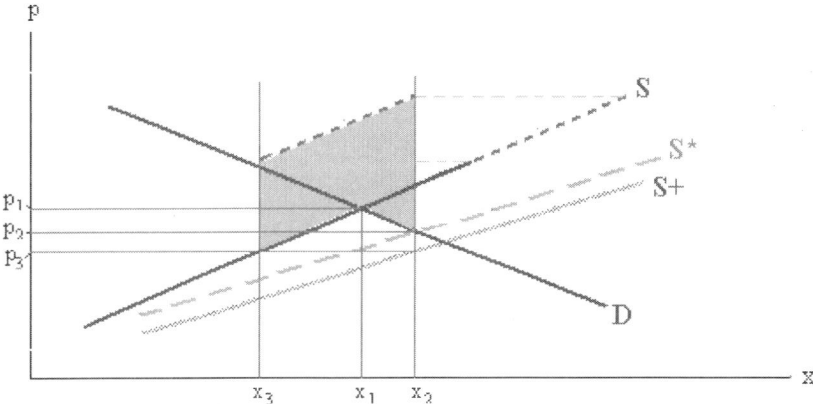

Figure 10.20. The loss economic surplus, shown as the gray shaded area, due to the implementation of the TGC market.

Note that the previous model assumes a closed market (autarchy), i.e. a Norwegian market, a Nordic market or a European market. The development may differ if export and import across the market boundaries are taken into account.

10.7.2 Conditions for a stable TGC market

Transactions on the TGC market should be sufficiently transparent in order to provide TGC price information. This information, together with price information from the electricity market, is decisive for potential investors upon deciding whether and when to invest in new renewable capacity. After the decision to invest, there is a time delay before new capacity actually enters the market, due to the time it takes to apply for and receive permits as well as the construction time. Once built, it is likely that the renewable electricity plants will be in operation, as most renewable technologies have low operational costs.

A TGC market will probably consist of a spot market and a forward market. Issued certificates will be traded on the spot market, while long-term contracts for future certificates will be traded on the forward market. The forward market allows hedging against price risks, thereby securing investments in renewable electricity projects.

10.8 References

[1] I. S. Kristensen: "Use of Green Sertificates in Nordic Power Supply. Analysis based on System Dynamic Modeling". Hovedoppgave Elkraftteknikk 2003.

[2] NVE: "Grønne sertifikater",Rapport 11, 2004.

[3] T. Bye, O. J. Olsen, K. Skytte: "Grønne sertifikater – design og funksjon" SSBrapport nr 11, 2002.

Abbreviations

AC	Alternating current
AFC	Average fixed cost
ATC	Average total cost
AVC	Average variable cost
BETTA	British Electricity Trading and Transmission Arrangement
CCGT	Combined cycle gas turbine
CEDB	Central Electricity Generating Board
CEER	Council of European Regulators
CHP	Combined heat and power
CSP	Curtailment service providers
DC	Direct current
DEGES	Director General of Electricity Regulation
EC	European Commission
EEX	European Energy Exchange (Germany)
EMCC	European Market Coupling Company
EMPS	EFI's multi area power scheduling
ERGEG	Council for European Regulators' Group for Electricity and Gas
ETSO	European Transmission System Operators
EU	European Union
FERC	Federal Electricity Regulatory Commission
FTR	Financial Transmission Rights
GNP	Gross national product
HHI	Hirshmann-Herfindahl index
HV	High voltage
HVDC	High voltage direct current
IEM	Integrated Electricity Market (in Europe)
IOU	Investor owned utility

Abbreviations

IPP	Independent power producer
ISO	Independent System Operator
ITC	Inter TSO compensation
LMC	Locational Marginal Cost
LMP	Locational Marginal Price
LOLP	Loss of load probability
LRMC	Long run marginal cost
MC	Marginal cost
MMC	Monopolies and Mergers Commission
NETA	New electricity trading arrangement
NGC	National Grid Company
NLDC	National Load Dispatch Centre (India)
NOK	Norwegian krone
NVE	Norsk vassdrags- og energiverk (Norwegian Water Resource and Energy Directorate
NZEM	New Zealand Electricity Market
N2EX	Spot Market UK
OFFER	Office of electricity regulation
OFGEM	Office of gas and electricity regulation
OPF	Optimal power flow
OTC	Over the counter
PAB	Pay-as-bid
PJM	Pennsylvania, New Jersey and Maryland
PPP	purchasing power parity
PURPA	Public Utility Regulatory Policy Act
PX	Power exchange
REC	Regional electricity company
SCDP	Security Constrained Economic Dispatch
SMC	System marginal cost
SO	System Operator

Abbreviations

SRMC	Short run marginal cost
STO	Regional Transmission Organization
TOU	Time of use
TPA	Third party access
TSO	Transmission system operator
VLL	Value of lost load